Plasma Etching Processes for Interconnect Realization in VLSI

Series Editor
Robert Baptist

Plasma Etching Processes for Interconnect Realization in VLSI

Edited by

Nicolas Posseme

First published 2015 in Great Britain and the United States by ISTE Press Ltd and Elsevier Ltd

ISTE Press Ltd
27-37 St George's Road
London SW19 4EU
UK

www.iste.co.uk

Elsevier Ltd
The Boulevard, Langford Lane
Kidlington, Oxford, OX5 1GB
UK

www.elsevier.com

British Library Cataloguing in Publication Data
A CIP record for this book is available from the British Library
Library of Congress Cataloging in Publication Data
A catalog record for this book is available from the Library of Congress
ISBN 978-1-78548-015-7

Printed and bound in the UK and US

Contents

List of Acronyms

ARC	Anti-reflective coating
ATRP	Terpinene
BARC	Bottom anti-reflective coating
BEOL	Back-end-of-line
CCP	Capacitively coupled plasma
CD	Critical dimension
CMOS	Complementary metal oxide semiconductor
CMP	Chemical–mechanical planarization
CVD	Chemical vapor deposition
DEMS	Diethoxymethylsilane
DFCCP	Dual frequency capacitively coupled plasma
DMDCS	Dichlorodimethylsilane
DS	Downstream
EM	Electromigration
ER	Etch rate
ESL	Etch stop layer
FC	Fluorocarbon
FDSOI	Fully depleted silicon on insulator

FEOL	Front-end-of-line
FSG	Fluorine-doped silicon glass
FTIR	Fourier transform infrared spectroscopy
HDMS	Hexamethyldisilazane
HF	Hydrofluoric acid
HSQ	Hydrogen silesquioxane
IC	Integrated circuit
ICP	Inductively coupled plasma
ITRS	International Technology Roadmap for Semiconductor
MEOL	Middle-end-of-line
MER	Mass etch rate
MERIE	Magnetically enhanced reactive ion etcher
MHM	Metal hard mask
MOSFET	Metal oxide semiconductor field effect transistor
MSQ	Methyl silsesquioxane
MTBC	Mean time between clean
OTMSA	Trimethylsilylacetate
P4	Post-porosity plasma protection
PECVD	Plasma enhanced chemical vapor deposition
PL	Planarizing layer
PPLK	Photopatternable low-k
PR	Photoresist
RC	Resistive-capacitive
RF	Radio frequency
RIE	Reactive ion etching
RMS	Root mean square

SEM	Scanning electron microscopy
SOI	Silicon on insulator
TCP	Transformer coupled plasma
TDDB	Time-dependent dielectric breakdown
TEOS	Tetra ethyl ortho silicate
TFMHM	Trench first metallic hard mask
TMCS	Trimethylchlorosilane
TMDS	Tetramethyldisilazane
TMSDEA	Trimethylsilydiethylamine
ULK	Ultra low-k
UV	Ultra violet
VLSI	Very large scale integrated
VUV	Vacuum Ultraviolet
XPS	X-ray photoelectron spectroscopy

Preface

Electronics and information systems play an ever-increasing role in the worldwide economy. With a global income of $265 billion in 2008, the semiconductor industry contributed to more than $1,300 billion in the electronics industry and $5,000 billion in services, which represented nearly 10% of the gross domestic product. Electronics and information systems have penetrated and transformed all aspects of life, including transportation, communications, health and well being, government services, banking systems, entertainment and education.

Micro- and nanoelectronics are the key enabling technologies for electronics, information and communications technology, and as a result, the semiconductor market is increasing at double the rate of gross domestic product growth. The specific position of the microelectronics industry has been made possible by the constant downscaling of device dimensions, which increases the performance to fulfill the current societal needs in consumer electronics that can now be divided into two main paths: performance improvement and energy efficiency. For this, during the last 50 years, integrated circuits have evolved from a 100-transistors chip in 1966 to multibillion-transistors circuits in 2010, with the smallest device measuring less than 20 nm.

However, the problem with the constant downscaling in dimension is that the resistive–capacitive (RC) delay coming from interconnects has become the main issue for devices with high performance. That is why copper-based interconnects have been introduced at the end of the 1990s to reduce the resistivity of wires, and low dielectric constant materials (low-k dielectrics) were introduced to deal with interconnects capacitance and signal propagation delay. But these changes have not been done without some difficulty.

Indeed, for the last 10 years, low-k dielectrics (i.e. materials with a lower dielectric constant than silicon dioxide) have evolved from fluorine-doped silicon glass (FSG; $k = 3.2$), to organosilicate (SiOCH; $k = 3.0$ and 2.7) and porous SiOCH ($k = 2.55$ and 2.4). Further decreasing the dielectric constant requires us to increase the porosity. Unfortunately, the presence of interconnected pores has amplified the dielectric sensitivity to plasma-based processes that are required for interconnects fabrication. It is thus impossible to define patterns in porous low-k materials without damaging their electrical properties with current etching technologies. Moreover, the mechanical strength of the dielectric, already significantly degraded by a decrease in network connectivity in SiOCH, is now reduced even further when porosity is added. For these reasons, technological roadmaps have constantly been revised and chip manufacturers are currently planning to step back to more robust dense materials with larger dielectric constant, thus degrading the global integrated circuits performance.

The goal of this book is to present the difficulties encountered for interconnect realization in very large-scale integrated (VLSI) circuits, especially focusing on plasma-dielectric surface interaction. After an introduction to interconnects presented in Chapter 1, we will see in Chapter 2 the sensitivity of low-k and ultra-low-k films to plasma etching and stripping steps which are the most critical steps in advanced interconnects realization. Then, in Chapter 3 we will present the various flows for dielectric films integration and their associated challenges. Finally, we will discuss in Chapter 4 the options to further reduce the dielectric constant for the future technological nodes.

Nicolas POSSEME

January 2015

1

Introduction

In 1947, the bipolar transistor was invented by Bardeen, Brattain and Shockley. Following this invention, in 1957, Kilby created five transistors simultaneously, forming the first integrated circuit (IC). In 1960, the first metal oxide semiconductor field effect transistor (MOSFET) on a silicon substrate with SiO_2 gate insulator was fabricated. The MOSFETs are slow compared to bipolar devices but are easier to fabricate and have a higher layout density. But both devices (bipolar and MOSFET) suffer from high power dissipation and have a restricted use in large integrated chip.

In 1963, the invention of the complementary metal oxide semiconductor (CMOS) marked a new milestone in the area of semiconductors. Indeed, the CMOS transistor has lower power dissipation and the possibility to integrate millions of CMOS transistors onto a chip. Since then, ICs have evolved from a 100-transistors chip in 1966 to multibillion transistors circuits in 2010, with the smallest device of less than 20 nm.

Semiconductor fabrication is composed of three major parts, front-end-of-line (FEOL), middle-end-of-line (MEOL) and back-end-of-line (BEOL), including different main steps such as deposition,

---

Chapter written by Nicolas POSSEME and Maxime DARNON.

lithography, etching and cleaning. The whole process flow represents several hundreds of steps for the manufacturing of chips.

The FEOL processes correspond to isolation, gate patterning, spacer, extension and source/drain implantation, silicide formation and dual stress liner formation.

The MEOL is mainly gate contact formation, which becomes more and more challenging as device dimensions are reduced [MEB 14].

The BEOL allows transistor functionality by electrically interconnecting transistors. Interconnects are composed of insulating layers (dielectric) and metal levels. Interconnects (see Figure 1.1) are composed of several metal levels. Each metal level is composed of horizontal metallic lines connected to the lower and upper metal levels through short vertical lines called vias.

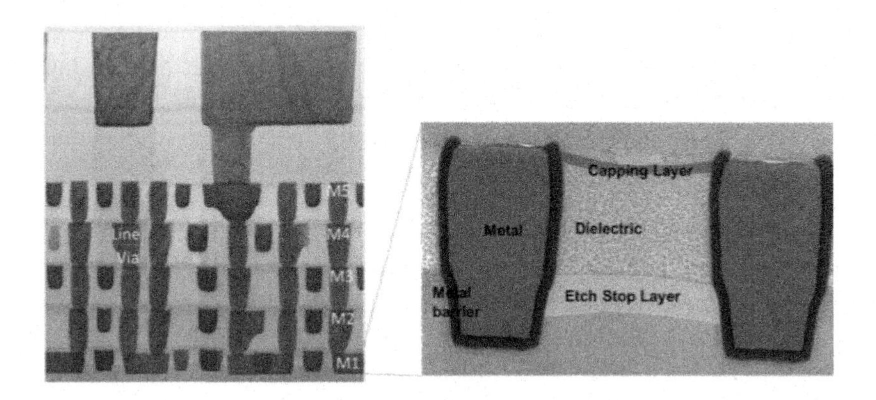

Figure 1.1. *Illustration of interconnects from metal 1 (M1) to metal 5 (M5) (left-hand side) and transverse cross section of two metal lines of a similar metal level (right-hand side)*

The combination of aluminum (Al) metal lines and silicon dioxide (SiO_2) as dielectric material has been used so far for interconnects in ICs. The problem lies with the constant downscaling in device dimensions resistive–capacitive (RC) delay (product of the resistance of the metal lines and their intercapacitance), dynamic power consumption ($\sim\alpha\ C.V^2f$ for a wire, where f is the frequency of digital signal, α is the wire activity factor, V corresponds to the voltage between the two digital levels and C is the total interconnect capacitance of a wire length) and cross-talk noise. These are becoming the main issues for devices with high performances [YAM 00, DEL 99].

Indeed, as shown in Figure 1.2, the reduction of device dimensions leads to an improvement of the device performances characterized by a decrease in the gate delay as a function of the generation. But from the 0.25 μm technology nodes, interconnects (SiO_2 + Al) become the limiting factor for high-performance devices leading to an increase in the total delay (see Figure 1.2).

Therefore, it is mandatory for BEOL interconnect wiring to be smaller, but also to cater for an increase in device density inducing important delay increases. Smaller dimensions require the introduction of new materials and new integration schemes.

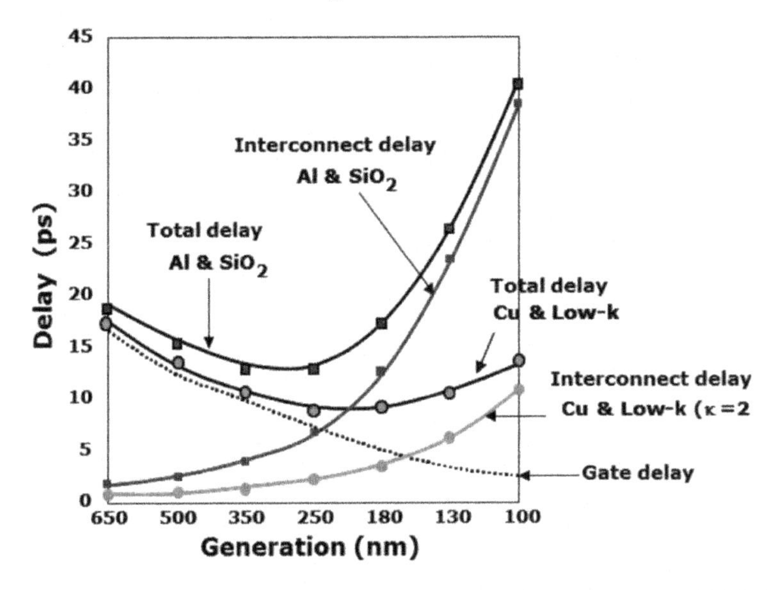

Figure 1.2. *Gate and interconnect delay versus technology generation [FUA 03]*

A model at the first order can be used to estimate interconnect RC delay as shown in Figure 1.3.

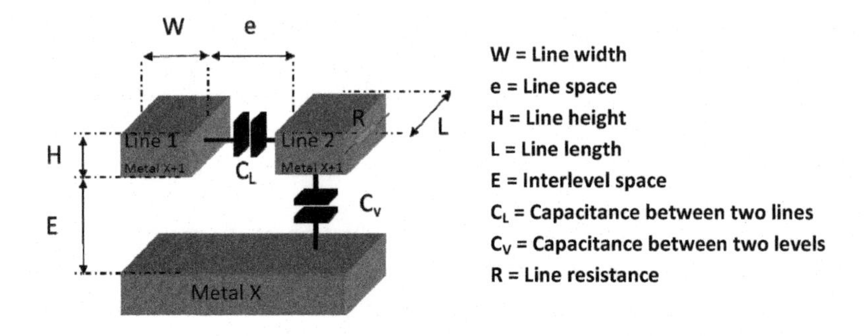

Figure 1.3. *Illustration of interconnect network*

The resistance of one interconnect line is given by:

$$R = \frac{\rho.L}{W.H} \quad [1.1]$$

where ρ is the metal resistivity, and L, W and H are the metal line length, width and height, respectively.

The level-to-level capacitance is defined by:

$$C_V = \frac{\varepsilon_0.k.L.W}{E} \quad [1.2]$$

where E and k are the dielectric film thickness between two levels and dielectric constant, respectively, and ε_0 is the permittivity of free space.

While the capacitance between the two interconnect lines is:

$$C_L = \frac{\varepsilon_0.k.L.H}{e} \quad [1.3]$$

where e is the space between two lines.

For this simple structure, the total line capacitance can be approximated by combining [1.2] and [1.3] as follows:

$$C = C_V + C_L = \varepsilon_0.k.\left(\frac{L.W}{E} + \frac{L.H}{e}\right) \qquad [1.4]$$

Related to [1.1] and [1.4], the RC delay defined as the product of the resistance of the metal lines and their intercapacitance is:

$$RC = \varepsilon_0.k.\rho L^2\left(\frac{1}{HE} + \frac{1}{we}\right) \qquad [1.5]$$

Therefore, related to [1.5] with the constant scaling down in dimension, the RC delay increases (see Figure 1.2). In addition, the dynamic power consumption for a wire and the cross-talk noise change with the capacitance and thus become larger when the dimensions shrink. As a result, interconnects are becoming the major limiter to device performance when dimensions are reduced.

To overcome this limitation, it was proposed to:

– replace the aluminum wiring with a metal presenting lower resistivity (ρ) [KEI 03];

– switch from conventional SiO_2 ($k \approx 4.2$) to an insulator with lower dielectric constant (low-k) [ROU 05, HOO 05, JOU 07a].

These changes marked a new milestone in the history of interconnects with the introduction of new integration processes and new materials.

1.1. Integration processes related to copper introduction

With a lower resistivity of about 30% compared to aluminum, copper was chosen as a new metal for interconnects [KEI 03].

Copper was introduced by IBM for the first time in a product in 1997 and since then, has widely been adopted by the semiconductor industry [VOL 10]. But switching from aluminum to copper implied to revisit the integration scheme in the BEOL. Indeed, copper is very difficult to etch using conventional plasma etching temperatures (40–60°C) due to the formation of non-volatile compounds [LEE 97, HOW 91, OHN 98]. This difficulty implied the modification of patterning and integration processes switching from metal patterning followed by dielectric filling to dielectric patterning followed by copper filling, as shown in Figure 1.4. This last approach is called damascene, referring to a method that begins by dielectric layer deposition, followed by etching structures (trenches or holes) into the dielectric.

To minimize the risk of metallic contamination by copper, a metallic barrier, usually made up of tantalum/tantalum nitride (Ta/TaN), is deposited before the copper. This step is then followed by the filling of structures with copper. A chemical–mechanical planarization (CMP) is finally used to remove the excess metal.

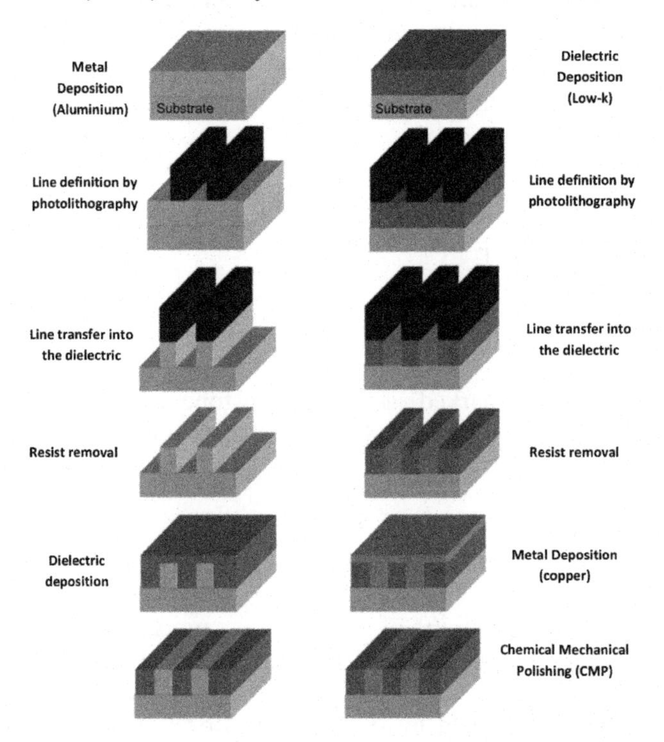

Figure 1.4. *Schematic of BEOL integration comparing conventional approach and damascene approach*

If only trenches or holes are fabricated, the method is called single damascene. In the dual damascene approach, both holes and trenches are etched into a dielectric followed by a metal fill and CMP. Since dual damascene forms at trenches and holes with one metal deposition and one CMP step, it is the favored approach. Dual damascene methods reduce the number of process steps, thus reducing fabrication costs.

1.2. Dielectric material with low-*k* value (<4)

The dielectric constant of materials is given by the Clausius–Mossotti equation as follows:

$$\frac{k-1}{k+2}=\frac{4}{3}\pi N\alpha \qquad [1.6]$$

where k is the dielectric constant of the material, N is the number of molecules per volume unit (density) in the material and α is its total polarizability, including electronic, ionic and dipolar polarizabilities:

– electronic polarization occurs in neutral atoms when an electric field displaces the nucleus with respect to the electrons that surround it;

– ionic polarization occurs when adjacent positive and negative ions stretch under an applied electric field;

– dipolar polarization occurs when permanent dipoles in asymmetric molecules respond to the applied electric field.

These different polarizations are presented in Figure 1.5. Therefore, according to [1.6], reducing the k-value of dielectrics is obtained by:

– reducing their polarizability by the use of low-polarity bonds (like C–C, C– H, Si–CH_3) [MAE 03];

– reducing their density by the introduction of porosity [MAE 03, JOU 07b].

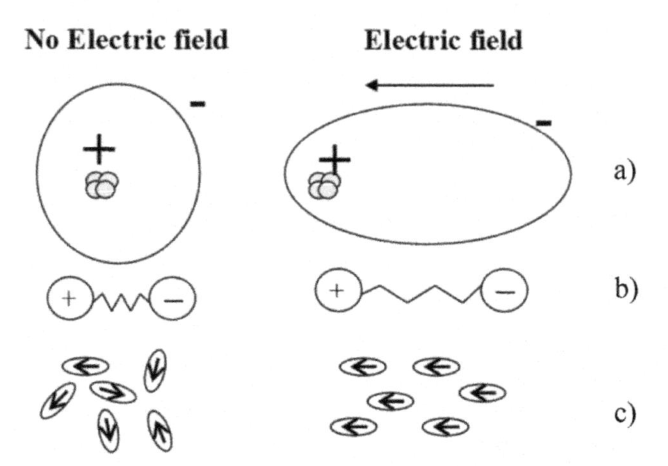

Figure 1.5. *Illustration of the three polarization mechanisms: a) electronic, b) ionic and c) dipolar*

Two categories of materials have been proposed to replace silicon dioxide: silicon-dioxide-based materials or polymers. These films can be dense or porous. They can be deposited by spin coating or by chemical vapor deposition (CVD). In all cases, these materials have a lower dielectric constant than SiO_2 and are called low-k materials [SHA 04].

In the spin-on deposition, the film coating is performed by dispensing a liquid precursor at the center of the substrate, which is placed on a spinner. Rotation of the substrate creates centrifugal forces that ensure an equal distribution of material on the surface. The thickness and uniformity of the coating is the result of the balance between centrifugal forces, viscous forces (determined by the viscosity of the solution) and evaporation rate of the solvent. The film is then heated to remove the solvent.

With the CVD approach, the deposition occurs due to reactions between gas precursors present in the gas phase. These reactions are thermally assisted with or without plasma. When plasma assists the deposition, the deposition is called plasma-enhanced CVD (PECVD).

1.2.1. *Organic low-k dielectrics*

These dielectrics are based on organic polymers. Saturated hydrocarbons have a lower polarizability than unsaturated, conjugated and aromatic hydrocarbons.

Therefore, they potentially provide the lowest *k*-value without requiring the introduction of porosity. Most of the organic low-*k* films have a dielectric constant in the range of 1.9–3.0. However, such films present integration issues. The major drawback is the low thermal and mechanical stability.

1.2.2. *Silicon dioxide-based films*

The first dielectric film used in the interconnects is silicon dioxide. This film is deposited by PECVD using SiH_4 + O_2 or tetraethyl orthosilicate (TEOS) + O_2. The dielectric constant of such a film is close to 4.

To reduce the dielectric constant of the film, some oxygen atoms can be replaced by fluorine hydrogen or CH_x groups.

1.2.2.1. *Fluorine-doped silicon glass film (FSG)*

The dielectric constant reduction is based on increased fluorine concentration in the dielectric film. Indeed, O-Si-F has less polar bonds than O-Si-O. However, the major drawback with a high amount of flourine in the material is the reaction of this element with water leading to the formation of hydrofluoric acid (HF) impacting the copper line integrity with copper corrosion [TRE 98, PAS 97]. Furthermore, different studies have suggested that to achieve stable low-*k* films, the amount of fluorine incorporated (atom %) should be less than 10% [SEN 97]. They demonstrated that the Young modulus and the hardness decrease with increasing fluorine incorporation. Despite these drawbacks, fluorosilicate glass (k = 3.8) was successfully implemented for the 130 nm technology node by ST Microelectronics.

1.2.2.2. *Hydrogen silsesquioxane (HSQ)*

For this film, the dielectric constant reduction is obtained by introducing hydrogen into SiO_2 film. The dielectric constant of such film is about 3, due to the lower film density.

1.2.2.3. *Methyl silsesquioxane (MSQ)*

The next step to further decrease the dielectric constant involves the addition of methyl groups ($-CH_3$). MSQ matrix materials have a lower dielectric constant (2.8–3) than HSQ. The addition of $Si–CH_3$ bonds introduces less polar bonds and, because of the larger size of the methyl group, also creates additional free volume compared to Si–H bonds.

The advantage of silicon-dioxide-based materials is that their chemical properties and structures are similar to those of traditional SiO_2, facilitating their integration in the BEOL. Such films were successfully integrated in the production line from the 90 nm technological node.

For instance, SiOCH (k = 3.1) film deposited by PECVD, BDI™ from Applied Materials, was successfully implemented at ST Microelectronics for the 90 nm technology node and then extended to the 65 nm.

1.2.3. *Porous films*

For advanced interconnect nodes, scaling down of the dielectric constant implies a continual decrease of the dielectric constant. This can be done by increasing the porosity and pore size in the material. In this case, low-k materials can be formed by PECVD through the co-deposition of a silica-like matrix together with a sacrificial organic polymer (porogen), the latter being subsequently removed by ultraviolet (UV)-assisted thermal curing at a temperature range of 300–450°C [JOU 07a], as shown in Figure 1.6. For spin-on deposition, the same approach can be used or the material can have a constitutive porosity using sol–gel processes.

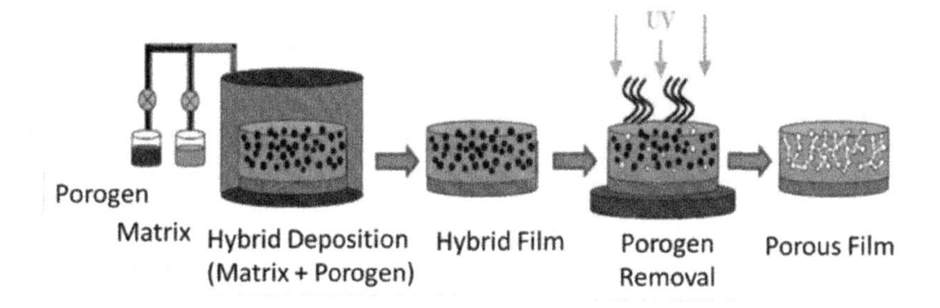

Figure 1.6. *Schematic representation of porous film deposition using a porogen approach*

The resulting low-*k* films are porous. The precise composition and porosity depends on the type of precursor molecules, the matrix/porogen ratio used during deposition and the curing conditions (typically heating at 400°C + broadband UV exposure for several minutes). The porous low-*k* films can be called mesoporous when the pore size diameter is greater than 2 nm or microporous when the pore size diameter is less than 2 nm.

For instance, microporous BD2x™ from Applied Materials, using diethoxymethylsilane (DEMS) and ATRP (terpinene) as matrix and porogen, respectively, presenting a dielectric constant of 2.5 with 25% porosity, was introduced for the 45 nm technology at ST Microelectronics.

Porous SiOCH film integration is very challenging. As an illustration, the International Technology Roadmap for Semiconductor (ITRS) (reference for the technological targets fixed by the industry) kept stepping back and delaying the integration of lower dielectric constant materials since 2000, as illustrated in Figure 1.7. Indeed, the ITRS edited in 2000 planned an introduction of low-*k* with dielectric constant of less than 1.2 in 2015. Today, the reality is that low-*k* films present a dielectric constant of approximately 2.8 (ITRS edited in 2013).

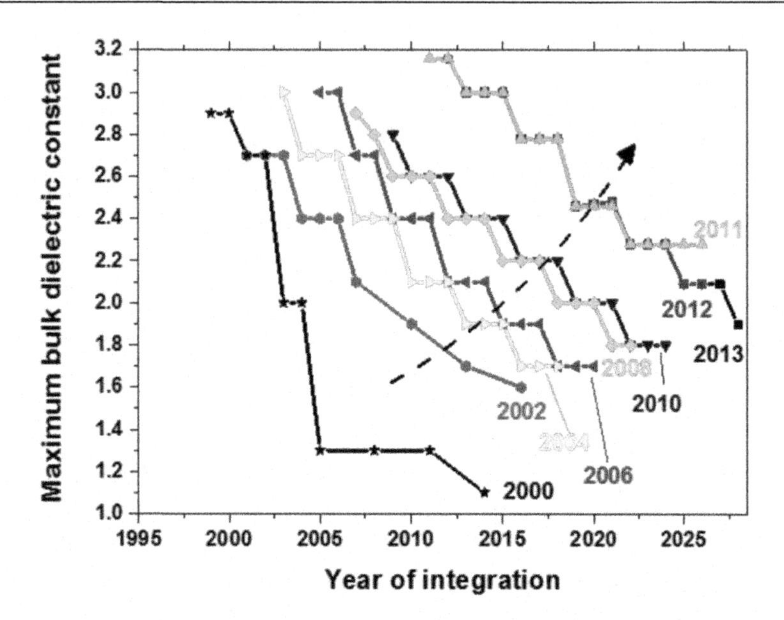

Figure 1.7. *Maximum effective dielectric constant per year of integration forecast by ITRS and summarized by the authors for the editions from 2000 to 2013*

The main requirements for a porous low-*k* material to be successfully integrated are:

– hydrophobicity: water presents highly polar O-H bonds with a *k* value close to 80. Even a small amount of absorbed water leads to an increase in the total *k* value. Therefore, low-*k* film must remain as hydrophobic as possible to prevent degradation of its dielectric constant;

– thermal stability: the low-*k* film must be compatible with the temperatures used for interconnect manufacturing (typically up to 400°C);

– mechanical stability: this condition is directly related to the use of damascene structures. The low-*k* film must be compatible with the mechanical strain imposed during the CMP step. In addition, cracks and delamination of multilevel interconnects structure can occur during the IC packaging if the low-*k* is not mechanically stable enough;

– stability under processing conditions: such as etching, cleaning, deposition and chemical mechanical polishing.

The last requirement is one of the most critical for porous low-*k* materials integration. Indeed, porous films are very sensitive to dry and wet processes. During dry processes, due to this porosity, plasma species can penetrate into the material and modify its structure (loss of methyl groups, moisture uptake, precursor diffusion), leading to a degradation of its dielectric performance. There are a lot of plasma processing steps encountered by low-*k* materials during the interconnect fabrication, like deposition of capping layers and dielectric masks on top of the porous material, dielectric etching (holes and trenches), cleaning (photoresist ashing) and deposition of copper anti-diffusion barrier.

In the next chapter, we present an overview of the interactions between plasma processes and porous low-*k* (SiOCH) materials. Then, we will present in more detail the various integration flows and their associated challenges. Finally, we will discuss the options to further reduce the dielectric constant for the future technological nodes.

2

Interaction Plasma/Dielectric

For the fabrication of interconnect structures, porous low-k dielectrics are exposed to various plasma processes during etching, during post-etch treatments used to remove carbon-based masks and during treatment surfaces. We will describe here the main plasma processes used for the integration of low-k films, focusing on plasma surface interaction for methylsilsesquioxane (MSQ) films (porous and non-porous SiOCH). When not mentioned, the film is deposited by plasma enhanced chemical vapor deposition (PECVD).

2.1. Porous SiOCH film etching

2.1.1. *Choice of etch tool*

A plasma is a partially ionized gas that is quasi-neutral. Electrons excited by an external power supply are heated to a temperature of few electron volts (eV). When they collide with neutral species, they can dissociate molecules, ionize the neutral species or excite the species to a higher energy level. Excited species emit photons when they go back to their fundamental level. As a result, a plasma is a complex mixture of ions and electrons (with an equal quantity of negative and positive charges), reactive radicals, stable molecules and energetic photons. The synergy between the ion bombardment and the

---

Chapter written by Nicolas POSSEME, Maxime DARNON, Thierry CHEVOLLEAU and Thibaut DAVID.

reactive neutrals favors reactions with the materials exposed to the plasma. Ion acceleration in the direction of the electric field boosts the process in the direction of the electric field, which leads to anisotropic processes. This effect is used for etching patterns with a controlled profile.

Plasmas used for etching in the semiconductor industry typically work at low pressure (few mTorr to hundreds of mTorr), with neutral species density several orders of magnitudes larger than charged species density, with an electron temperature of a few eV and ion energy from a few tens of eV to few hundreds of eV.

There has been a serious evolution of the etching reactors for dielectrics in the past 30 years to meet the demand of etching processes (see Figure 2.1).

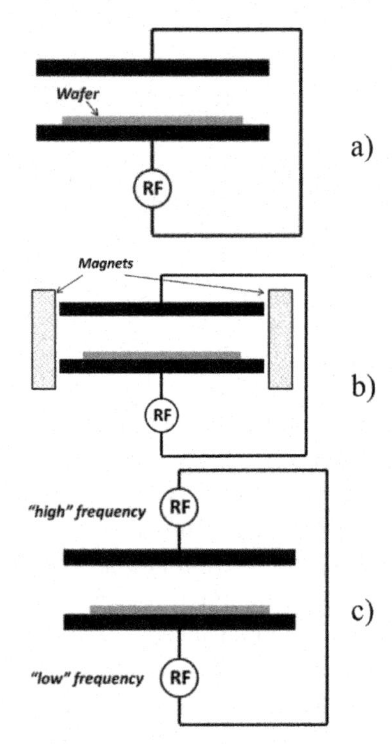

Figure 2.1. *Presentation of dielectric etch tool developed in the past 30 years: a) capacitively coupled plasma (CCP), b) magnetically enhanced reactive ion etcher and c) dual frequency CCP*

The first generation of dielectric etch tool presented a low-density capacitively coupled plasma (CCP) (Figure 2.1(a)).

In the CCP etch tool, parallel plate electrodes are excited by a radio frequency (RF) power supply. This creates an alternating electric field perpendicular to the electrodes. Electrons are accelerated by this electric field and can then ionize and dissociate the molecules of the plasma. This electric field can also accelerate the ions toward the electrodes, which makes plasma formation (dissociation and ionization) and ion acceleration (sheath potential) directly coupled. In CCP, the ion energy can reach several hundred eV. The plasma pressure must be between 20 and 300 mT for the electrons to be able to ionize the molecules. The resulting plasma density is low, between 10^9 and 10^{10} electon.cm^{-3}. However, increasing the power not only increases the plasma density, but also the self-bias voltage (Vdc) and thus the ions' energy. As a result, plasma density and ion energy are coupled and cannot be changed independently.

Inductively coupled plasma (ICP) sources which alleviate these problems were introduced in the 1990s. They provide high ion flux of relatively low-energy ions to the wafer which is useful for several applications such as silicon and metal etching. In ICP, the plasma density and ion energy are controlled separately. The energy is transferred to electrons through inductive coupling by a coil. The plasma density can be easily increased (10^{11}–10^{12} electrons.cm^{-3}) without increasing ion energy. In order to control the ion energy, a second generator is capacitively coupled to the substrate. The ion energy can reach tens of eV and a few hundred eV. However, for SiO_2 etching, a high-energy ion bombardment is required. For this, magnetically enhanced reactive ion etcher (MERIE) was developed (Figure 2.1(b)).

MERIE etch tools are similar to CCP but allow higher plasma density (10^{10} cm^{-3}) due to a magnetic electrons confinement. A magnetic field is added and is used to prevent electron neutralization on the chamber walls. Therefore, the electron lifetime in the plasma increases and the plasma density increases. The plasma generated is non-uniform due to the drift imposed by the magnetic field ($v \times B$ where v is the electron velocity and B is the local magnetic field). Therefore, electrons accumulate on one side of the wafer leading to a strong Vdc non-uniformity across the wafer.

This issue can be addressed by using rotating magnetic fields to homogenize the plasma density above the wafer. A rotating magnetic field is made by sequentially exciting the electromagnets around the chamber, or by physically rotating the magnets.

Finally, the necessity to get simultaneously energetic bombardment and with high ion flux with independent control leads to the development of dual frequency CCP (DFCCP) (Figure 2.1(c)). In CCP plasma, the efficiency of electron excitation or ion acceleration depends on the excitation frequency. As shown in Figure 2.2, lower frequencies lead to high ion energy and low plasma density (illustrated here by the ion flux), while higher frequencies lead to lower ion energy and higher plasma density. Multi-frequency capacitive plasmas take advantage of this effect by using a high-frequency power supply to control the ion density and obtain medium-density plasmas and low-frequency power supply to control the ion bombardment energy. Current industrial tools use two or three power supplies with frequencies ranging from a few MHz to 100–200 MHz.

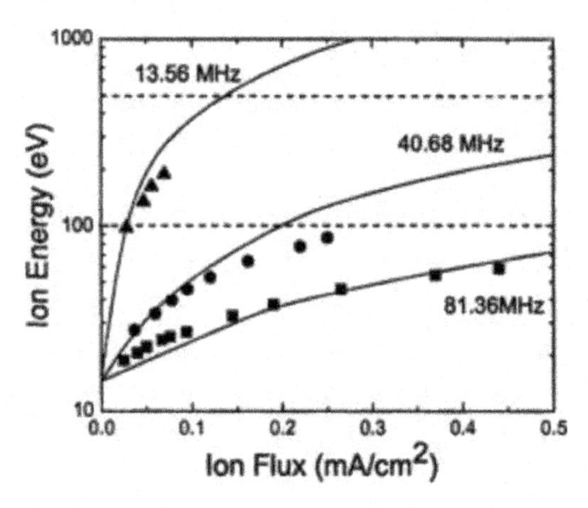

Figure 2.2. *Ion energy versus ion flux in an argon capacitive plasma [PER 05]*

2.1.2. *Dielectric film etching*

SiOCH material etching requires halogen-based plasmas to remove Si. The most common halogen used is F that forms volatile SiF_x compounds. Other halogens can be used for low-k etching, but there is the risk of generating metal corrosion [BAK 13a]. Therefore, most etching processes use fluorocarbons (C_xF_z) or fluorohydrocarbon gases ($C_xH_yF_z$).

Fluorine forms volatile products with silicon, carbon and hydrogen, while carbon forms volatile products with oxygen and hydrogen. However, the etching process is efficient only in the presence of ion bombardment that provides enough energy to break the Si–O and Si–C bonds.

As previously mentioned, the SiOCH film was chosen for its similarities with SiO_2. The etching of SiO_2 has widely been studied in the literature [CAR 90, STA 99]. It was shown that the etch rate of SiO_2 in fluorocarbon-based plasmas is governed by a thin fluorocarbon layer that forms on the surface. The thickness and composition of this layer are defined by plasma conditions (pressure, power and composition) as well as by the composition of the material itself (oxygen that is released during SiO_2 etching reacts with the fluorocarbon layer on the surface).

Do we have similar etch mechanisms with SiOCH films?

2.1.2.1. *Etch mechanism of SiOCH films*

An etch mechanism comparison between SiO_2 and SiOCH (BDI^{TM}) films was made using fluorocarbon-based plasmas ($CF_4/N_2/Ar$-based process) performed in MERIE. The etch mechanism has been understood on blanket wafer.

Increasing the polymerizing rate in the gas phase (by adding C_4F_6 or CH_2F_2 to $CF_4/N_2/Ar$) leads to decreasing etch rate for both films (Figure 2.3(a)).

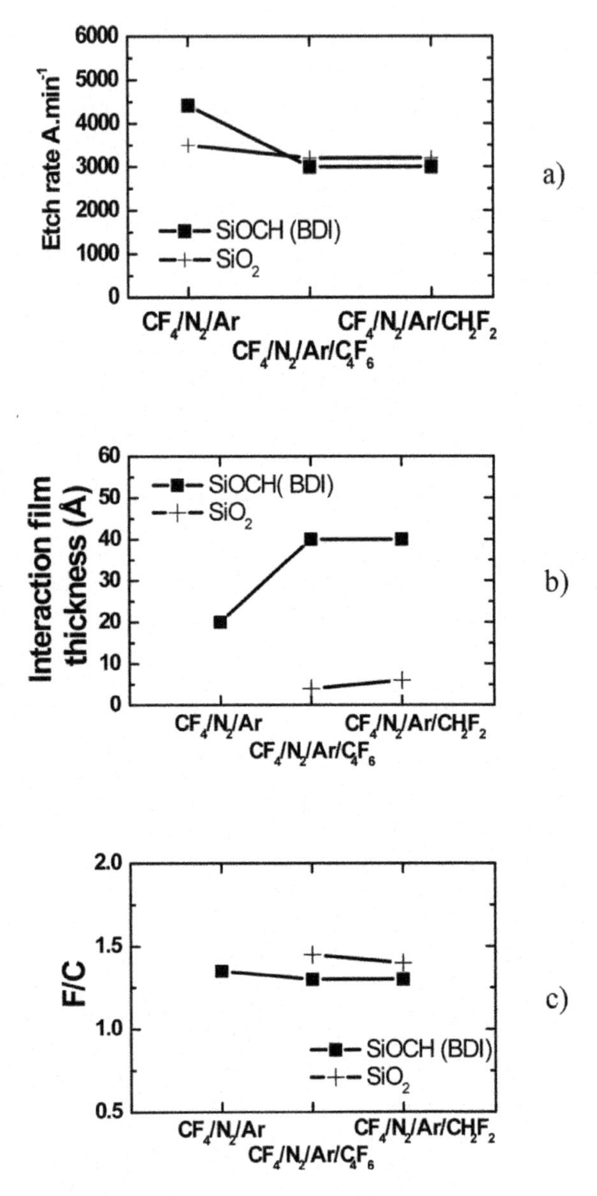

Figure 2.3. *Etch rate evolution of SiOCH and SiO_2 exposed to fluorocarbon-based plasma a) correlated with perturbed layer thickness b) and its composition ([F]/[C] ratio) c)*

Fourier transform infrared spectroscopy (FTIR) analysis, giving information on SiOCH film structure change after partial etching, shows in Figure 2.4 that the as-deposited spectrum exhibits a main shoulder at 1,034 cm^{-1} corresponding to the Si-O-Si stretching vibration mode. Two peaks are also observed at 1,275 and 2,960 cm^{-1}, which are assigned to the Si-CH_3 and C-H_x vibration mode, respectively.

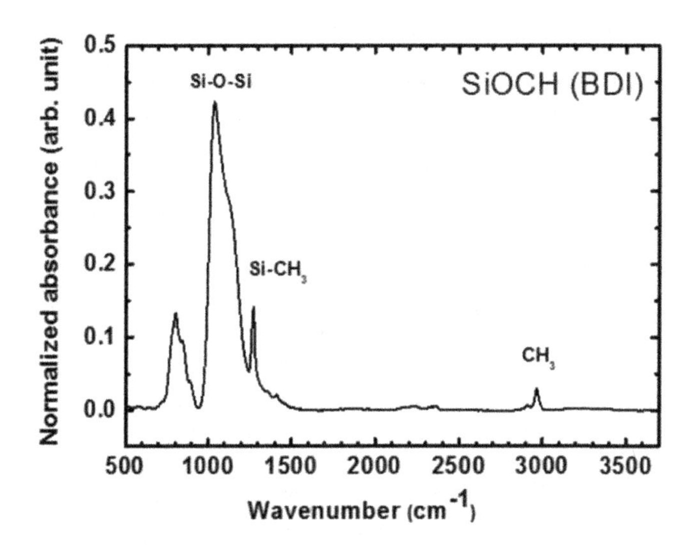

Figure 2.4. *As-deposited SiOCH FTIR spectrum*

After SiOCH partial etching using $CF_4/N_2/Ar$, FTIR spectrum is similar to the pristine film (not shown here). The normalized absorbance to the thickness is the same and no other vibration bands are observed, indicating that the remaining SiOCH material is not altered by the fluorocarbon etching process according to FTIR sensitivity.

Angle-resolved X-ray photoelectron spectroscopy (XPS) analyses, giving information on the surface composition, reveal in Figure 2.5 that the fluorine and carbon concentrations increase when the takeoff angle increases while the amount of silicon and oxygen decreases. This result indicates that the fluorocarbon concentration increases when the analyzed depth decreases (larger takeoff angle) demonstrating that the fluorocarbon layer is localized at the top surface of SiOCH.

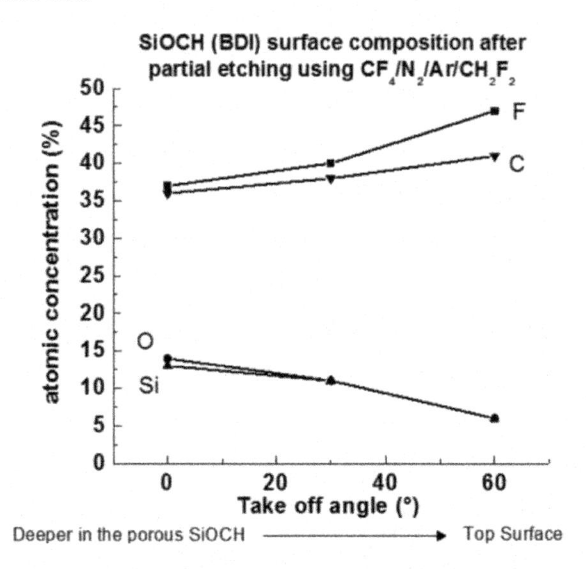

Figure 2.5. *SiOCH surface composition versus takeoff angle (with respect to the normal) after partial etching in $CF_4/N_2/Ar/CH_2F_2$*

Due to XPS, a correlation between the fluorocarbon layer thickness and composition formed on top of the film during etching has been done (Figures 2.3(b) and (c)).

Indeed, the thickness can be extracted from Si2p XPS spectrum using the following well-known expression:

$$d_{lay} = \lambda_{CF} \cos(\theta) \ln\left(\frac{I_{Si}^{\infty}}{I_{Si}}\right) \qquad [2.1]$$

where λ_{CF} is the escape depth of silicon photoelectrons in the perturbed layer, θ is the takeoff angle of photoelectrons with respect to the normal to the surface (equal to 45°), I_{Si}^{∞} is the Si2p core-level intensity measured on the as-deposited low-k without the fluorocarbon film and I_{Si} is the Si2p core-level intensity measured with the fluorocarbon film after etching.

While the composition of the fluorocarbon layer can be quantified from C1s spectrum by calculating the [F]/[C] ratio, which is obtained from the C1s spectra using the following equation:

$$[F]/[C] = \frac{3 \times I_{CF3} + 2 \times I_{CF2} + I_{CF}}{I_C} \qquad [2.2]$$

where I_{CF3}, I_{CF2} and I_{CF} are the integrated peak areas of the CF_x groups (x = 1, 2, 3) in the C1s spectrum and I_C is the total integrated area of the C1s peaks originating from the fluorocarbon layer.

More details about this experimental setup can be found elsewhere [POS 03].

These measurements can be made based on a bilayer model making the assumption that a perturbed layer is formed on top of the investigated materials. This is confirmed by angle-resolved XPS analyses and infrared spectroscopy (FTIR) previously described.

Based on XPS spectra obtained after etching, it was shown that the etching using highly polymerizing process is controlled by the fluorocarbon film formed at the upper surface for both films. Indeed, with polymerizing gas addition to $CF_4/N_2/Ar$ plasmas, important slowdown in etch rate of the different films investigated here is observed (Figure 2.3(a)). This is correlated with an increase in the perturbed layer thickness (Figure 2.3(b)), while its compositions ([F]/[C]) are similar whatever the use of the polymerizing gas (Figure 2.3(c)). In this case, when the fluorocarbon layer thickness increases, less ion energy is dissipated in the bulk dielectric material, thereby inducing a slowdown of the etch rate (Figures 2.3(a) and (b)).

But, the fluorine content in the fluorocarbon layer also has an impact on the etch rate. Indeed, an increase in the Ar concentration in the CF4/N2 gas mixture leads to a decrease in etch rate

(Figure 2.6(a)). This behavior is correlated with a lower fluorine content in the perturbed layer (Figure 2.6(b)), while its thickness remains similar whatever the argon dilution (Figure 2.6(c)). In this case, a decrease in fluorine content in the fluorocarbon layer induces a lower etch rate due to less free fluorine available to reach the dielectric interface.

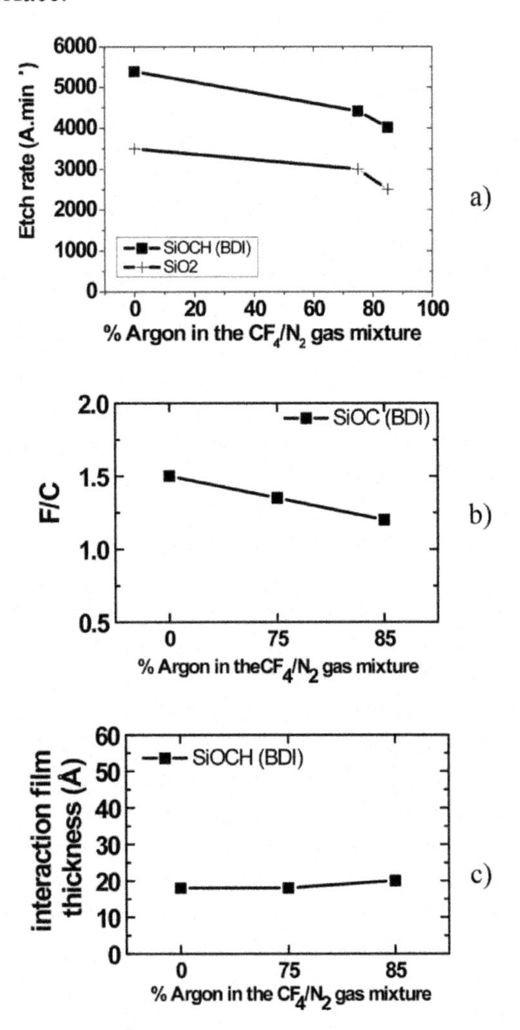

Figure 2.6. *Etch rate evolution of SiOCH and SiO_2 with respect to argon dilution in the CF_4/N_2 gas mixture a) correlated with perturbed layer b) and its composition ([F]/[C] ratio) c)*

Finally, the chemical composition of the dielectric material also has a strong impact on the formation of the fluorocarbon layer (thickness and composition). The oxygen concentration in the film induces thinner fluorocarbon layer and higher [F]/[C]. Indeed, the less oxygen present in the dielectric, the thicker the fluorocarbon film formed on top of the dielectric during plasma etching, as illustrated in Figure 2.3. The more oxygen present in a dielectric, the more fluorinated is the perturbed layer.

Therefore, these results show that when SiO_2 is replaced by a SiOCH material, similar etch mechanisms are observed with the formation of the fluorocarbon layer that controls the etching. The thickness and composition of the fluorocarbon layer are influenced by the material composition and plasma parameters.

2.1.2.2. *Etch and modification mechanism of porous SiOCH film*

In this section, the etching of porous SiOCH films (presenting different degrees of porosity 50, 40 and 30% with 2–3 nm pore size) in medium-density fluorocarbon plasmas has been investigated. All the materials investigated are spin coated and exhibit a dielectric constant between 2.1 and 2.3. The study has been performed on blanket films.

Similarly to non-porous SiOCH (BDI film previously described), the etch rate of porous materials in CF_4/Ar-based process decreases with higher Ar dilution or polymerizing gas addition such as CH_2F_2 (Figure 2.7). The etch rate increases when the degree of porosity in the SiOCH film increases, indicating that a porous film etches faster than a non-porous film. This can be explained by less material per unit thickness that needs to be removed when the porosity increases.

Figure 2.8 shows the FTIR spectrum of the as-deposited film with 50% void. We observe a main absorption band at 1,056 cm^{-1} corresponding to the Si-O-Si stretching vibration mode and two additional peaks at 1,275 and 2,960 cm^{-1} assigned to the Si-CH_3 and C-H_3 vibration mode, respectively. Similar FTIR spectra are observed for the other porous SiOCH (30 and 40%). Based on these FTIR reference spectra, all the as-deposited porous SiOCH investigated consist of a siloxane (Si-O-Si) network terminated by methyl groups ($Si-CH_3$).

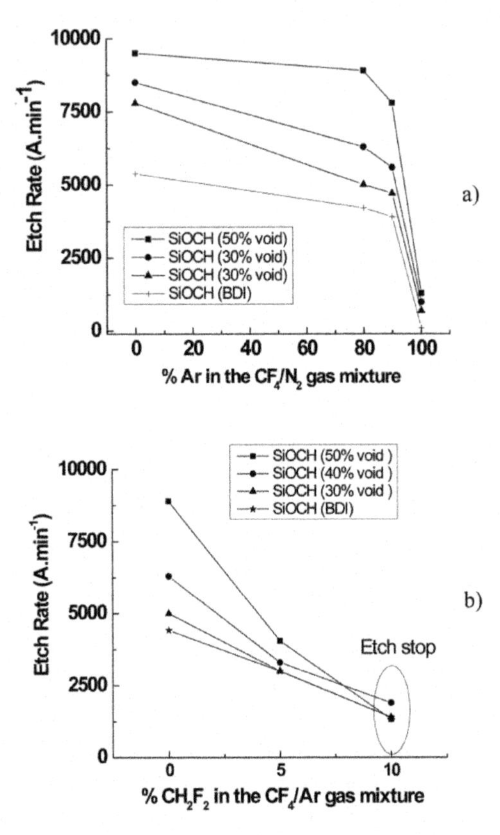

Figure 2.7. *Etch rate evolution of porous SiOCH with respect to argon dilution in the CF_4/N_2 gas mixture a) or CH_2F_2 addition to CF_4/Ar gas mixture b)*

After partial etching using fluorocarbon-based process, FTIR spectra (not shown) exhibit the same absorption bands than the pristine porous SiOCH films. However, a change in peak position and intensity is observed for the Si-O-Si and $Si-CH_3$ peaks, respectively. This result indicates that the remaining bulk material is altered by the etching process.

From these FTIR analyses, the relative $Si-CH_3$ content can be determined by comparing the peak height ratio between the $Si-CH_3$

and Si-O-Si stretching vibration modes [GRI 03]. More details about this experimental protocol can be found elsewhere [POS 04].

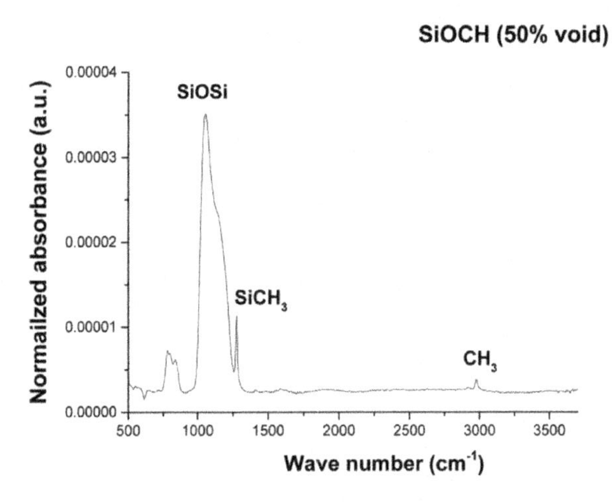

Figure 2.8. *As-deposited SiOCH (50% void) FTIR spectrum*

Based on this peak comparison, Figure 2.9 shows the more porous the material is, the more the film degradation is important. Therefore, contrary to non-porous SiOCH, porous materials are altered during exposure to medium-density fluorocarbon-based plasmas [POS 04].

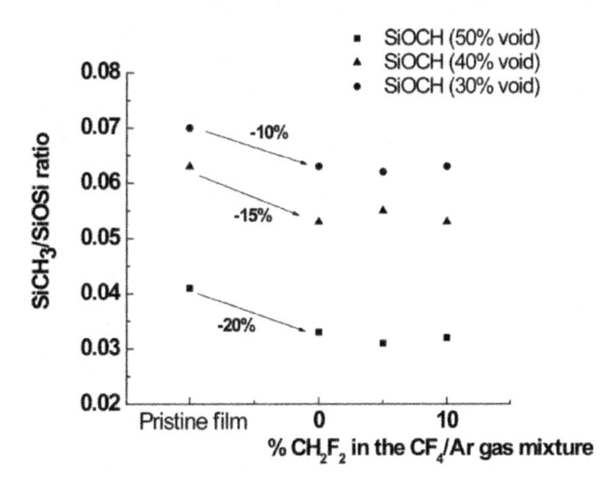

Figure 2.9. *Porosity impact on SiOCH film modification ($SiCH_3$ consumption) after partial etching in CF_4/Ar-based plasma versus CH_2F_2 content addition*

XPS analyses show no difference in surface composition as a function of the takeoff angle (Figure 2.10). This trend is observed whatever the porous film or etching process investigated, indicating that a mixed reactive layer containing Si, O, C, H and F is formed during porous low-k etching. Therefore, contrary to non-porous film, a mixed reactive layer is formed during etching and attributed to the diffusion of fluorocarbon species throughout the pores into the porous SiOCH film.

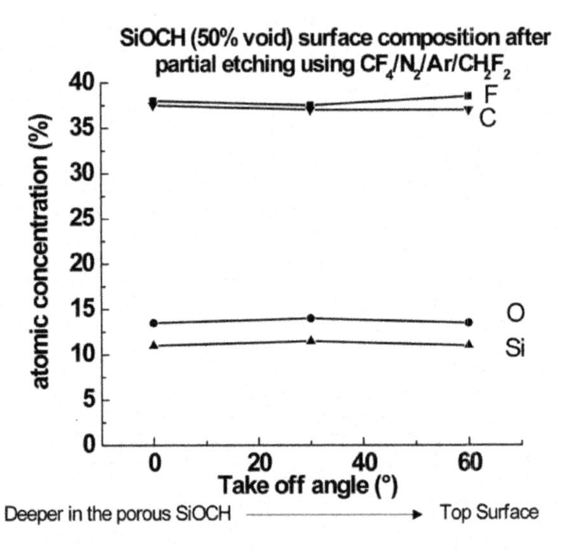

Figure 2.10. *Porous SiOCH (50% void) surface composition versus takeoff angle (with respect to the normal) after partial etching in $CF_4/N_2/Ar/CH_2F_2$*

2.1.2.2.1. Physical and chemical components' impact on porous SiOCH film modification

To discriminate the role of physical ion bombardment and chemical species diffusion on porous SiOCH degradation, the methyl group consumption has been measured after porous SiOCH (with void variation from 30 to 50%) exposure to purely physical Ar plasma and to a highly chemical inductive SF_6 plasma process. We demonstrated that in both cases the porous films are strongly damaged by physical or chemical plasma [POS 04].

In the first case, the high-energy Argon ions break Si-O, Si-C and C-H bonds, releasing H and O that can diffuse inside the pores and degrade the porous SiOCH film. This release of hydrogen is also an internal source of process that enhances the sputtering rate expected during Ar plasma exposure. This allows us to explain the high porous SiOCH etch rate (more than 100 $nm.min^{-1}$, see Figure 2.9) in pure argon compared to non-porous SiOCH.

In the second case, using SF_6 plasma (without bias), the fluorine deeply diffuses in the film and reacts with the material by forming Si–F bonds and releasing methyl groups. Fluorine reactive species are also very reactive toward porous SiOCH since such process etches the dielectric film with an etch rate of 300 $nm.min^{-1}$ for porous SiOCH with 50% void.

2.1.2.2.2. Effect of polymerizing process

Different mechanisms are observed in fluorocarbon plasmas. In this case, the carbon depletion is much less important for the different porous SiOCH films than in pure Ar or SF_6 plasmas, indicating a lower degradation of the porous materials after partial fluorocarbon etching. This relatively low degradation of porous material observed during etching in fluorocarbon-based plasmas is attributed to:

– the removal of the damaged film as it forms by the high chemical sputtering rate of the fluorocarbon plasma (more than 500 $nm.min^{-1}$ compared to only 100 $nm.min^{-1}$ using only argon (see Figure 2.7));

–the limitation of fluorine diffusion and absorption of ion energy by the mixed SiOCHF layer (demonstrated in Figure 2.10) formed at the surface of the porous material during plasma exposure.

2.1.2.2.3. Effect of porosity

Porous films are expected to be etched faster than non-porous films since less material per thickness unit is removed in a porous material.

Standaert *et al.* [STA 00] studied the etching of various low-k films with different pore sizes and porosities, assuming that the etch rate should scale according to the porosity of the material, following a simple law:

$$ER_{norm} = ER\,(1 - p) \qquad [2.3]$$

where ER_{norm} is the normalized etch rate in $nm.min^{-1}$, ER is the etch rate in $nm.min^{-1}$ and p is the porosity of the SiOCH materials. Another solution to account for the porosity would be to use the mass etch rate [DAR 13a], defined by the mass of material etched per second. In this case, the mass etch rate is calculated by:

$$MER = A \times \rho \times ER \times 10^{-4} \qquad [2.4]$$

where MER is the mass etch rate in $mg.min^{-1}$, ρ is the material density (in $g.\ cm^{-3}$) and A is the etched area in cm^2.

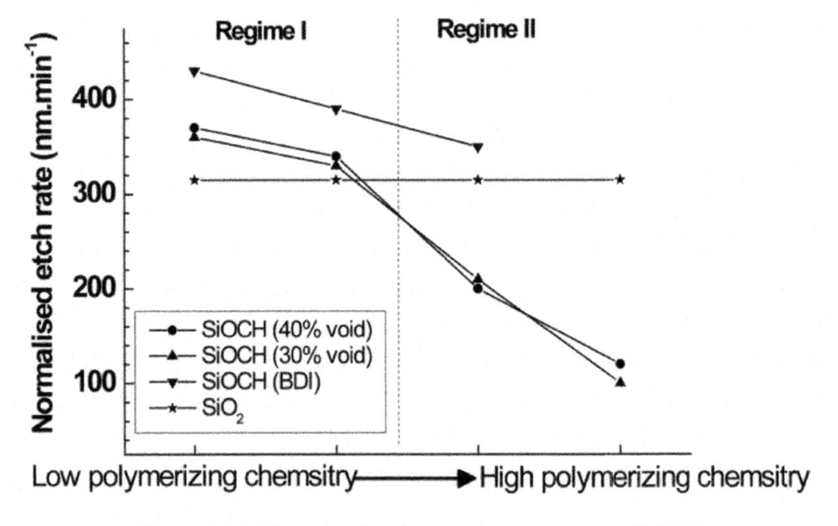

Figure 2.11. *Normalized etch rate of various porous SiOCH (presenting similar composition) and non-porous SiOCH using different etch chemistries from low to high polymerizing*

Based on equation [2.3], Figure 2.11 compares the normalized etch rate of different porous materials (30 and 40% porosity) with the etch rate of a non-porous SiOCH. These films present similar composition, therefore, we can directly see the impact of the porosity on the etch rate.

When the plasma is hardly polymerizing (CF_4/Ar gas mixtures) (Regime I), the normalized etch rate of porous SiOCH with 30 and 40% porosity in the film is identical. The normalized etch rate of both materials is significantly lower than the etch rate of the dense SiOCH materials with CF_4/Ar chemistries but higher than the etch rate of SiO_2.

These results clearly show that the porosity induces a slowing down in etch rate even in low polymerizing chemistries. The addition of a highly polymerizing gas, such as CH_2F_2 in CF_4/Ar (Regime II), strongly reduces the etch rate for the different porous films investigated and even stops the etching on blanket surfaces for the higher CH_2F_2 flow. We can also note that, under the same conditions, the etching does not stop on non-porous SiOCH, also demonstrating that the etch stop phenomenon is attributed to the presence of pores in the material.

We can conclude that the presence of porosity induces an accumulation of the fluorocarbon species at the porous material surface and can lead to the growth of a thick fluorocarbon polymer. The etching process of porous SiOCH film is a trade-off between the fluorocarbon radical diffusion rate through the mixed layer formed at the surface (porous SiOCH mixed with fluorocarbon) and the chemical sputtering rate of the mixed layer. If the fluorocarbon radical diffusion rate is faster than the chemical sputtering rate of the mixed layer, then the fluorocarbon polymer concentration at the etch front increases as a function of time and can lead to conditions where the etching stops with a smooth surface. On the contrary, if the fluorocarbon radical diffusion rate is lower than the chemical sputtering rate of this mixed layer, then the etching proceeds and the surface is rougher.

2.1.2.2.4. Summary

The factors responsible for porous SiOCH film damage have been identified as ion bombardment and reactive species diffusion (hydrogen and fluorine). Therefore, the current solution to limit the damage of porous SiOCH film is to control the formation of a protective layer, limiting effect of ion bombardment and fluorine species diffusion during etching in fluorocarbon-based plasma. In this case, a trade-off has to be found between the fluorocarbon radical diffusion rate through the mixed layer formed at the surface and the chemical sputtering rate of the mixed layer. A summary of the porous SiOCH film damage mechanisms during etching is presented in Figure 2.12.

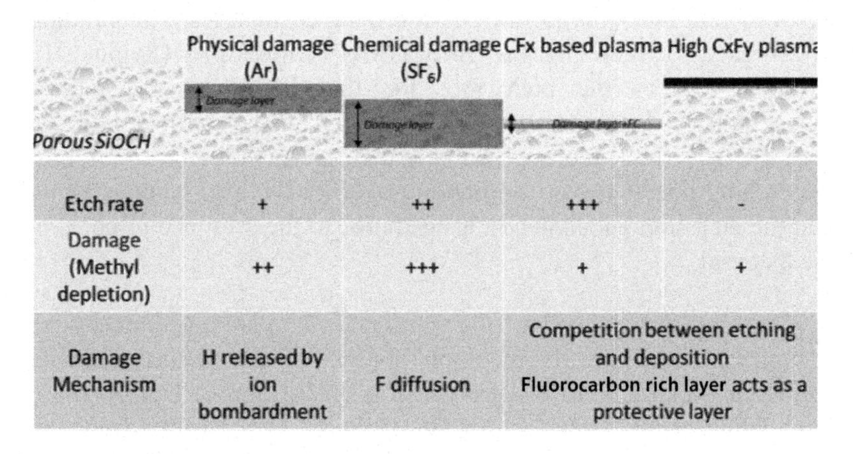

Figure 2.12. *Summary of the porous SiOCH film damage mechanisms (from low (+) to high (+++)) as a function of the plasma etch conditions*

2.1.2.3. *Surface roughness after etching*

Different studies have shown that porosity is responsible for porous SiOCH roughening during etching [LAZ 05, YIN 07, YIN 08, GAL 89].

Tatsumi *et al.* [TAT 05] suggested that low-k materials are roughened when a thin fluorocarbon layer (4 nm) is deposited on the material surface, while no fluorocarbon or thicker fluorocarbon layers do not lead to roughening. Yin *et al.* [YIN 07, YIN 06] attributed the etching-induced roughness to fluorocarbon species micromasking the material surface. They attributed the non-uniform coverage of the surface to the presence of pores in the material. Lazzeri *et al.* [LAZ 05] proposed that the larger surface of porous materials leads to an incomplete coverage of the surface of the dielectric by the fluorocarbon layer, compared to dense dielectric layers. Due to this incomplete coverage, plasma species can directly interact with the low-k material, leading to harsher etch conditions and roughness. In both cases, the models assume that the roughness is created by an incomplete or non-homogeneous coverage of the porous SiOCH surface by fluorocarbon species.

In a recent study, we demonstrated that the roughness of porous SiOCH ($BD2_x$™) also occurs under fluorocarbon-free plasma exposure (SF_6 plasma performed in ICP etch tool), suggesting complementary mechanism [BAI 10].

The mechanism of surface roughening consists of the following steps and shown in Figure 2.13:

1) surface densification: during the first few seconds of the etching process, the surface of porous SiOCH materials gets denser;

2) crack formation: cracks are formed, leading to the formation of deep and narrow pits;

3) radical diffusion: plasma radicals diffuse through those pits and the pore network and modify the porous material at the bottom of the pits;

4) roughness propagation and amplification: the difference in material density and composition between the surface and the bottom of the pits leads to a difference in etch rate and an amplification of the roughness.

The solution to prevent porous SiOCH roughening during etching is to use highly polymerizing Fluorocarbon (FC) chemistries. The impact of the etch process on porous SiOCH roughening is illustrated in Figure 2.14 by comparing the root mean square (RMS) roughness on blanket wafers induced by two different fluorocarbon-based etching chemistries (different degrees of polymerizing) performed in the MERIE etch tool. CH_2F_2 addition to CF_4/Ar (more polymerizing process) reduces the final roughness.

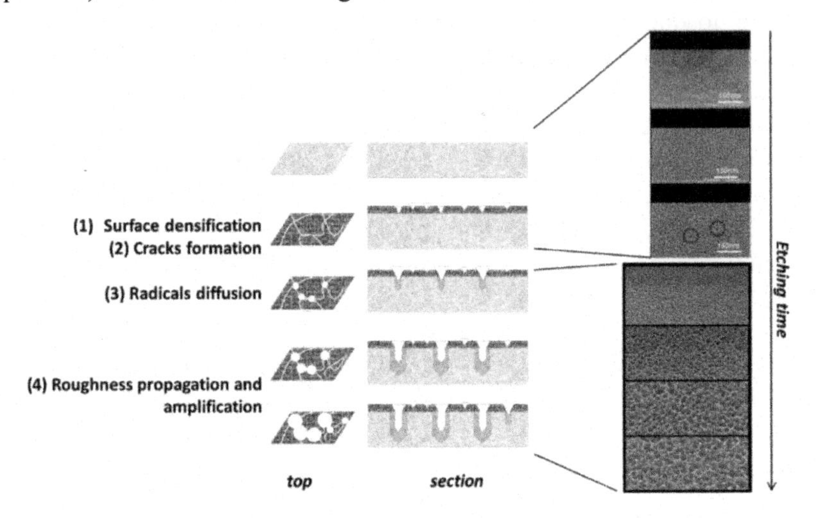

Figure 2.13. *Schematic evolution of porous SiOCH surface evolution during the etching process*

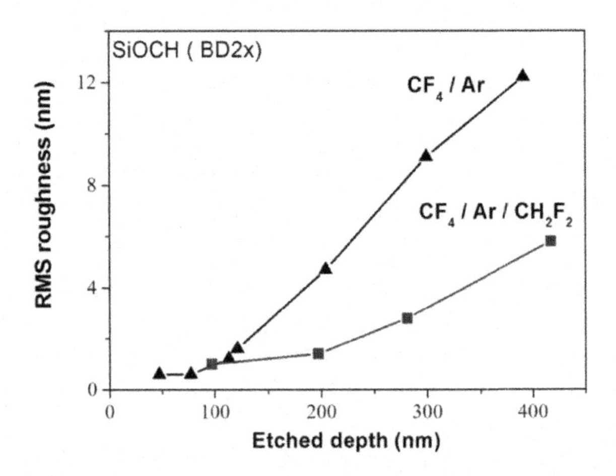

Figure 2.14. *Roughness evolution as a function of the etched depth for PECVD porous SiOCH (25% porosity, k = 2.5) exposed to either CF_4/Ar or CF_4/Ar/CH_2F_2 plasmas performed in MERIE. RMS value of the roughness measured by AFM*

We have previously shown that the porous SiOCH etching process development relies on a trade-off between the fluorocarbon radical diffusion rate through the mixed layer formed at the surface and the chemical sputtering rate of the mixed layer. This trade-off should also take into account the surface roughening that is also mitigated by the use of polymerizing etching processes.

2.1.2.4. *Pattern etching*

The mechanisms described above and performed on blanket wafers correspond to the etch front during pattern etching. However, pattern sidewalls are not exposed to high-energy ions during the process and different mechanisms can take place. In addition, the damaged layer formed during the etching process (and not fully removed by wet cleaning, depending on the process) increases the k-value of the dielectric between two adjacent trenches.

Figure 2.15 shows patterns etched in a dense or porous dielectric films (deposited by spin coating) after etching in fluorocarbon process ($CF_4/Ar/CH_2F_2$) performed in MERIE using in this case a carbon-based mask. The plasma-induced damage on sidewalls of patterns has been revealed using the hydrofluoric acid (HF) decoration method. This method is based on the fact that a pristine SiOCH is not etched by an HF dip, the opposite of a damaged film where the carbon depletion converts the material into a hydrophilic SiO_2-like one that is quickly attacked by HF. As a result, after dipping into an HF solution, cross-sectional scanning electron microscopy (SEM) inspection reveals which part of the SiOCH film is converted into a SiO_2-like material and removed by HF [DAR 10]. The space between the dielectric patterns is filled by a polymer before dipping the sample in HF to improve the SEM observation.

Based on this experimental protocol, Figure 2.15 shows the film degradation at the sidewalls directly scales with the degree of porosity in the material. Based on the etch/modification mechanism previously described, this film damage can be assigned to fluorine diffusion through the pores.

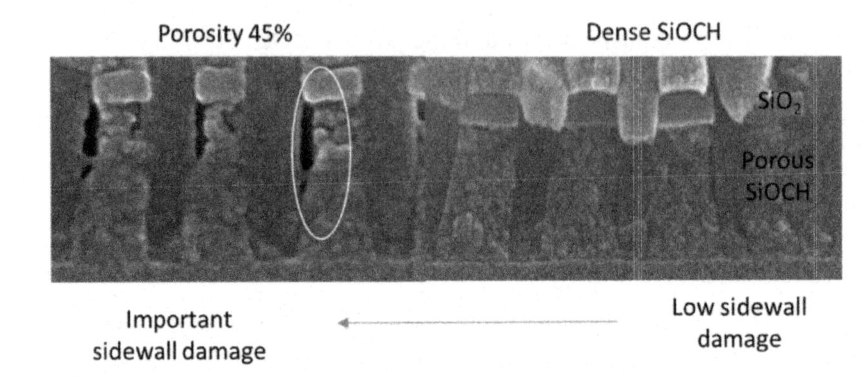

Figure 2.15. *Effect of porosity in the spin-on SiOCH film on sidewall film damage after line etching in $CF_4/Ar/CH_2F_2$ performed in MERIE*

One way to limit this sidewall film damage is to use a high polymerizing process. Indeed, using a polymerizing process, a thick CFx protective layer is formed at the sidewalls and it limits the species diffusion, preventing the porous film damage. The benefit of this process is illustrated in Figure 2.16 showing that the sidewall porous SiOCH (45% void) damage is strongly reduced with high polymerizing gas (CH_2F_2) in addition to CF_4/Ar process. However, a trade-off has to be found between the use of a high polymerizing process limiting the sidewalls damage of patterns and the risk to stop the etching.

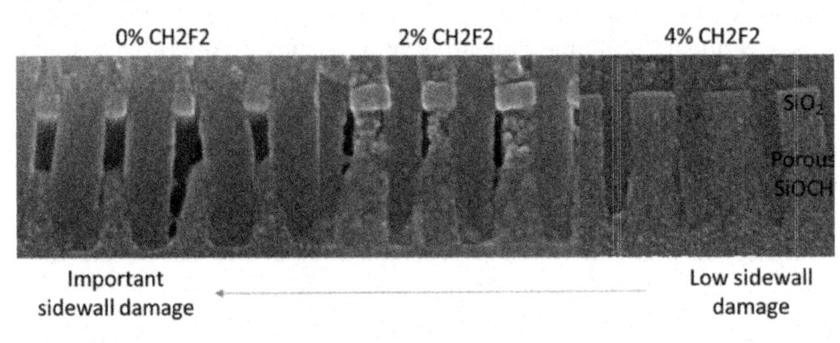

Figure 2.16. *Evolution of sidewall film modification (using decoration method) of porous SiOCH (containing 45% porosity) after etching as a function of the CH_2F_2 concentration in CF_4/Ar gas mixture*

2.2. Porous SiOCH film sensitivity to post-etch treatments

After the etching process, the porous low-k structures may be exposed to various kinds of post-etch plasma processes used to remove the carbon-based mask, decrease the fluorine contamination and prevent the formation of residues on the mask, etc.

These plasma processes can be divided into two categories: oxidizing plasmas that contain oxygen and reducing plasmas that contain hydrogen [LOU 04, POS 07]. These plasmas can be realized in reactive ion etching (RIE) or the downstream (DS) chamber. Contrary to the RIE etch chamber, in the DS reactors, only the dissociated molecules or atoms can reach the surface of the wafer. However, the absence of charged species means that no anisotropic processing is possible; therefore, such reactors can only be used for isotropic processes such as photoresist stripping or surface treatments.

A summary of the efficiency of such chemistries to remove an organic mask in our experimental conditions is presented in Figure 2.17. We can see that despite their efficiency in removing the organic mask, the different processes investigated also induce porous SiOCH (spin on, with 50% void) film damage.

	Oxydizing		Reducing	
	Downstream	RIE	Downstream	RIE
Efficiency to remove PR	++	+++	+	+++
Damage	++++	+	+	++
Damage Mechanism	Carbon depletion and Si-OH formation			

Figure 2.17. *Efficiency of reducing and oxidizing chemistries to remove the organic hard mask as well as their compatibility with porous SiOCH (50% void) film damage (from low (+) to high (++++)) estimated using the decoration method*

The mechanisms of porous low-k modification induced by an oxidizing or reducing process are now discussed. The experiments have been performed on blanket wafers.

2.2.1. *Oxidizing process*

When an SiOCH film is exposed to an O_2 plasma, an SiO_xH_y layer is formed at the dielectric film surface. Oxygen reactive species can deeply diffuse into the film through its porosity and react at the pores surface, forming dangling bonds or silanol (Si–OH) terminations. Dangling bonds can be converted into silanol groups upon moist air exposure. The formation of silanol groups inside the material makes it hygroscopic, leading to water condensation into the pores when the wafer is stored in the clean room after plasma exposure [DAR 13b]. The following mechanisms have been proposed in the literature [POS 07, CHA 02, WOR 05] to explain the film modification:

$$\equiv Si-CH_3+4O \rightarrow \equiv Si-OH+CO_2+H_2O \quad \Delta Hr1=-2266 \text{ kJ.mol-1 at 298 K} \quad [2.5]$$

$$\equiv Si-OH+HO-Si\equiv \leftrightarrow \equiv Si-O-Si\equiv +H_2O \quad \Delta Hr2=0 \text{ kJ.mol-1 at 298 K} \quad [2.6]$$

ΔHri corresponding to the enthalpy of the reaction (i).

Reaction [2.5] shows that the reaction of oxygen with porous low-k can extract methyl groups, and produce Si–OH terminations. Volatile carbon dioxide and water are produced by the reaction. Silanol groups can either remain in the material and induce moisture absorption or condense to reform SiOSi (reaction [2.10]). Based on these two reactions, the reaction of oxygen with porous SiOCH film leads to a methyl-depleted film with silanol terminations.

These reactions are confirmed by FTIR analyses comparing the structural composition of blanket porous SiOCH exposed to oxidizing chemistries performed in RIE or DS mode (Figure 2.18). After exposure to oxidizing chemistries in both DS and RIE plasma modes, FTIR spectra (see Figure 2.18) exhibit similar vibration modes as the pristine material one. But, additional absorption bands are observed between 3,000 and 3,700 cm^{-1}. These bonds are assigned to isolated and associated hydroxyl O–H and water groups which indicate moisture uptake. The presence of hydroxyl groups is also confirmed by a new absorption peak localized at 960 cm^{-1} that is attributed to SiO–H bonds (mainly observed in DS-O_2). Important film damage (methyl loss and moisture uptake) is observed when the porous SiOCH is exposed to DS-O_2 process (no methyl groups are observed). A lower film depth modification has been detected when the porous SiOCH film is exposed to RIE-O_2 plasma rather than to DS-O_2 plasmas. However, the impact of O_2 plasmas performed in DS or RIE mode is difficult to compare since the plasma conditions are strongly different (reactive species concentrations, ion bombardment energy and wafer temperature).

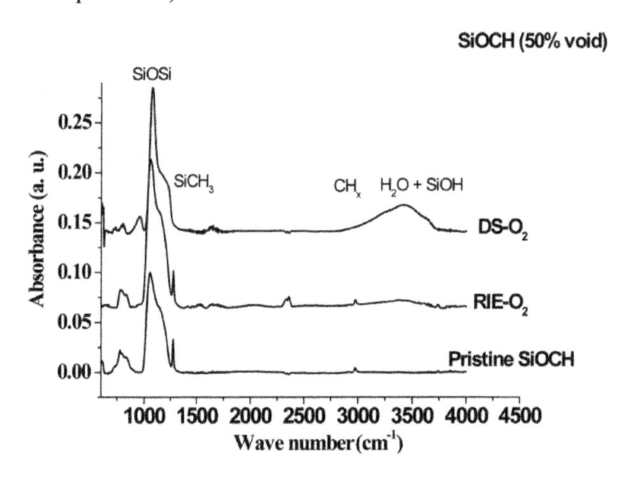

Figure 2.18. *FTIR spectra of porous SiOCH (spin on, 50% void) after exposure to oxidizing process performed in RIE or downstream (DS) mode*

Whatever the choice of O_2 plasma process, O_2 plasmas are prohibited for the porous SiOCH film integration.

2.2.2. *Reducing process*

Contrary to oxidizing plasmas, hydrogen species are thermodynamically less reactive with porous SiOCH than oxygen radicals. To operate, the reaction needs energy coming from ion bombardment and/or temperature and/or energetic photons. The possible reactions between hydrogen and porous SiOCH films are the following:

$$\equiv Si-CH_3+N+3H \rightarrow \equiv Si-NH_2+CH_4 \quad \Delta Hr3=-418 \text{ kJ.mol}^{-1} \text{ at } 298 \text{ K} \quad [2.7]$$

$$\equiv Si-O-Si\equiv+N+3H \rightarrow \equiv Si-NH_2+\equiv Si-OH \quad \Delta Hr4=+42 \text{ kJ.mol}^{-1} \text{ at } 298 \text{ K} \quad [2.8]$$

$$\equiv Si-NH_2+H_2O \rightarrow \equiv Si-OH+NH_3 \quad \Delta Hr5=+42 \text{ kJ.mol}^{-1} \text{ at } 298 \text{ K} \quad [2.9]$$

These reactions involve the formation of H_2O and CH_4 volatile by-products. Hydrogen reacts with Si–O–Si and Si–CH_3 to form Si–H and Si–OH terminating groups that can lead to a structural change of the SiOCH matrix. The enthalpy of reaction [2.7] is much lower than in reactions [2.8] and [2.9] indicating that H atoms break more easily Si–CH_3 bonds than Si–O bonds to preferentially form Si–H bonds.

During the subsequent exposure to the moist environment, Si–H groups can be hydrolyzed and can form silanol terminations through the following the equations:

$$\equiv Si-H+H_2O \rightarrow \equiv Si-OH+H_2 \quad \Delta Hr6=-478 \text{ kJ.mol}^{-1} \text{ at } 298 \text{ K} \quad [2.10]$$

This film degradation in reducing process is amplified when nitrogen is added in the gas mixture (for instance, NH_3, CH_4/N_2 and H_2/N_2).

During NH_3 plasma exposure, Si–OH and $SiONH_2$ groups are formed and methyl groups are consumed. The role of nitrogen is explained by the following mechanisms:

$$\equiv Si-CH_3+N+3H \rightarrow \equiv Si-NH_2+CH_4 \quad \Delta Hr7=-1375 \text{ kJ.mol}^{-1} \text{ at } 298 \text{ K} \quad [2.11]$$

$$\equiv Si-O-Si\equiv+N+3H \rightarrow \equiv Si-NH_2+\equiv Si-OH \quad \Delta Hr8=-915 \text{KJ.mol}^{-1} \text{ at } 298 \text{ K} \quad [2.12]$$

Once again, the material is further modified when it is exposed to the moist ambient. Si–NH_2 groups are replaced by Si–OH during the reaction:

$$\equiv Si-NH_2+H_2O \rightarrow \equiv Si-OH+NH_3 \quad \Delta Hr9=-258 \text{ kJ.mol}^{-1} \text{ at } 298 \text{ K} \quad [2.13]$$

Recent studies also showed that ultraviolet (UV) and vacuum ultraviolet (VUV) radiations impact the SiOCH film modification [UCH 08, LEE 10]. Energetic UV–VUV radiation can break Si–C bonds and Si–O. Therefore, UV–VUV radiation also plays a role in the SiOCH film modification.

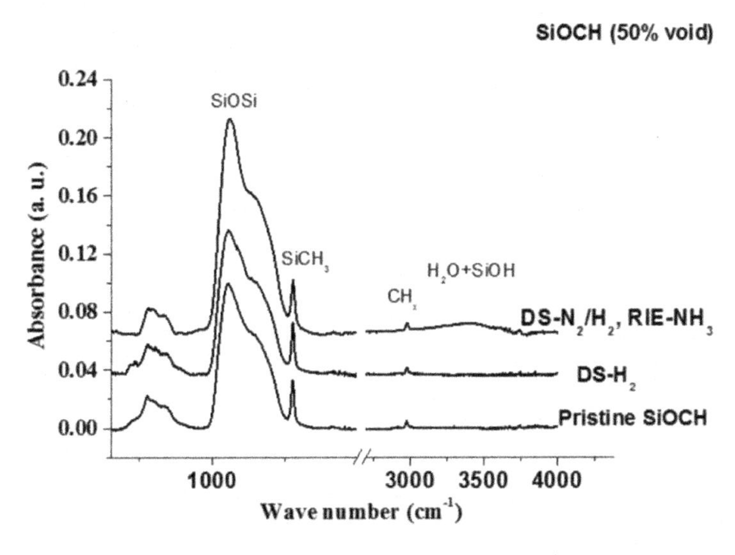

Figure 2.19. *FTIR spectra of porous SiOCH (spin on, 50% void) after exposure to reducing process (with or without nitrogen addition) performed in RIE or downstream mode*

The impact of reducing chemistries (as well as the nitrogen effect) is confirmed by FTIR analyses comparing the structural composition of blanket porous SiOCH exposed to reducing chemistries with or without nitrogen addition performed in RIE or DS mode (Figure 2.19). Figure 2.19 shows that the normalized absorbance of the porous SiOCH after DS-H_2 plasma exposures is the same as the pristine film one (no carbon depletion and no other vibration bands are detected).

When nitrogen is added to the hydrogen-based process, FTIR spectra reveal a new absorption band between 3,000 and 3,700 cm^{-1} assigned to isolated and associated hydroxyl OH and water groups, confirming that the film degradation in reducing process is amplified when nitrogen is added into the gas mixture.

2.2.3. *Origin of the k-value degradation*

Based on the previous work relating to the impact of oxidizing and reducing impact on porous SiOCH film damage, we have plotted the evolution of the k-value, the carbon depletion and the moisture uptake on porous SiOCH (spin on, 50% void) after DS-N_2/H_2, RIE-O_2 and RIE-NH_3. A comparison with CH_4 process performed in ICP etch tool has also been done to clarify the impact of the methyl group consumption on k-value decorrelated from the moisture uptake. The description of the experimental detail can be found elsewhere [POS 07].

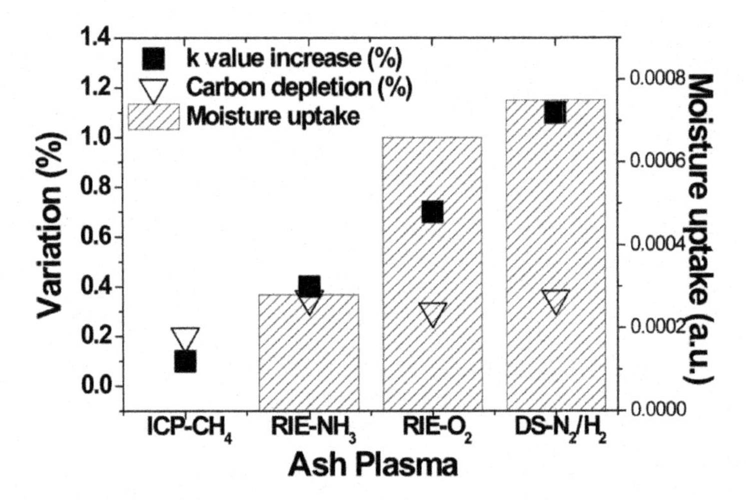

Figure 2.20. *Impact of the carbon depletion and moisture uptake on k-value variation for porous SiOCH (spin on, 50% void) as a function of ash plasma*

Under our experimental conditions, we have shown that, after the ICP-CH_4 process, despite a 20% loss of the methyl groups, the k-value is slightly impacted (+10%) (Figure 2.20). This behavior is attributed to and correlated with the absence of moisture absorption in the remaining porous material. Figure 2.20 illustrates that the increase in k-value is mainly related to moisture uptake rather than carbon depletion, irrespective of the plasma treatments. After DS-N_2/H_2 ash plasma exposure, Figure 2.21 shows higher moisture uptake and carbon depletion than in RIE chemistries. Both carbon consumption and moisture uptake lead to a strong increase in k-value (+50%). After RIE ash chemistries (NH_3 and O_2), the carbon depletion is identical for both chemistries, while a higher water uptake is detected after RIE-O_2 plasma exposure compared to RIE-NH_3 process. In this case, a higher k-value is detected for the RIE-O_2 plasma.

This result highlights that the increase in the k-value is mainly correlated with the amount of moisture uptake, and less correlated with the carbon depletion.

We have shown in this chapter that with porosity introduction, the dielectric film presents a higher sensitivity upon fluorocarbon-based plasma chemistries and ashing plasma exposures (oxidizing or reducing) which can lead to dielectric constant increase.

During the etching step, the factors responsible for porous SiOCH film damage have been identified as ion bombardment and reactive species diffusion (hydrogen and fluorine). The solution to limiting this sidewall film damage is to use high polymerizing process, protecting the sidewalls by the formation of a thick CF_x layer. But in this case, a trade-off has to be found between this sidewall protection and the risk to stop the etching.

During the ashing step, the ability of reactive species diffusion through the pore network leads to specific mechanisms depending on the choice of the ash process and results in methyl group consumption and moisture uptake, strongly increasing the dielectric constant.

We will now describe more precisely the process architectures and process flows used for pattern formation and will discuss the advantages and drawbacks of the various integration alternatives.

3

Porous SiOCH Film Integration

In complementary metal oxide semiconductor (CMOS) technology, the integration of porous low-*k* materials becomes mandatory from the 45 nm technological node and beyond in order to reach interconnect performances. As previously described, such porous low-*k* materials can easily be damaged by all the process steps (especially etching and post-etch treatment processes) that can induce a potential increase in the dielectric constant. In this context, different hard mask strategies such as metallic or organic masks have been proposed for porous SiOCH patterning. Each integration scheme presenting advantages and drawbacks for porous SiOCH film integration is dealt with in this chapter.

3.1. Trench first metallic hard mask integration

The trench first metallic hard mask (TFMHM) scheme proposed for low-*k* film introduction is presented in Figure 3.1.

In this TFMHM approach, a thin metallic hard mask layer (usually a titanium nitride layer (TiN)) is deposited on the top of a dense dielectric layer, which serves to encapsulate the underlying low-*k* material (Figure 3.1(a)).

Chapter written by Nicolas POSSEME, Maxime DARNON, Thibaut DAVID and Thierry CHEVOLLEAU.

In this TFMHM description, if not mentioned, the low-*k* film presented is a porous SiOCH deposited by plasma-enhanced chemical vapor deposition (PECVD) with a 25% porosity and a dielectric constant of 2.5 (BD2x^{TM}).

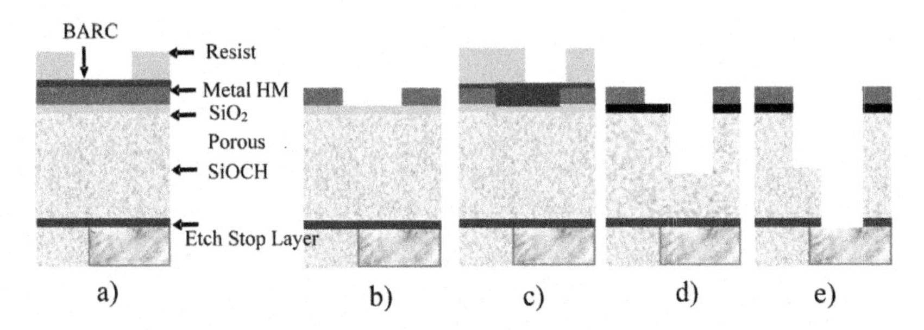

Figure 3.1. *TFHM architecture sequence: a) line lithography using an organic BARC, b) HM etching, c) via BARC/lithography, d) via etching and resist removal and e) line etching and etch stop layer opening*

3.1.1. *Metal hard mask opening*

The first step for photoresist (PR) pattern transfer into the TiN layer is the bottom anti-reflective coating (BARC) (in which its primary benefits in photolithography are focus/exposure latitude improvement, enhanced critical dimension (CD) control, elimination of reflective notching and protection of resist from substrate poisoning) opening followed by the metal hard mask opening (Figure 3.2(b)). Etching of the TiN film can be achieved by using radicals such as F, CF_x, H, Cl and BCl_x generated from the plasma [TON 03].

The study by Tonotani *et al.* [TON 03] showed that for TiN material etching, Cl_2 addition to argon gas is most effective for obtaining a high etching rate, and that BCl_3 or CHF_3 addition enables control of the taper angle of the etched profile. Indeed, although F or Cl radicals generated in the plasma enhance the etching rate of TiN by forming volatile compounds such as TiF_4, NF_x, $TiCl_x$ and NCl_x, non-volatile compounds such as TiF_3 or BO_xN_y are produced in the case of CHF_3 or BCl_3 addition.

Figure 3.2(b) illustrates the metal hard mask opening using a typical $Ar/Cl_2/BCl_3$ chemistry.

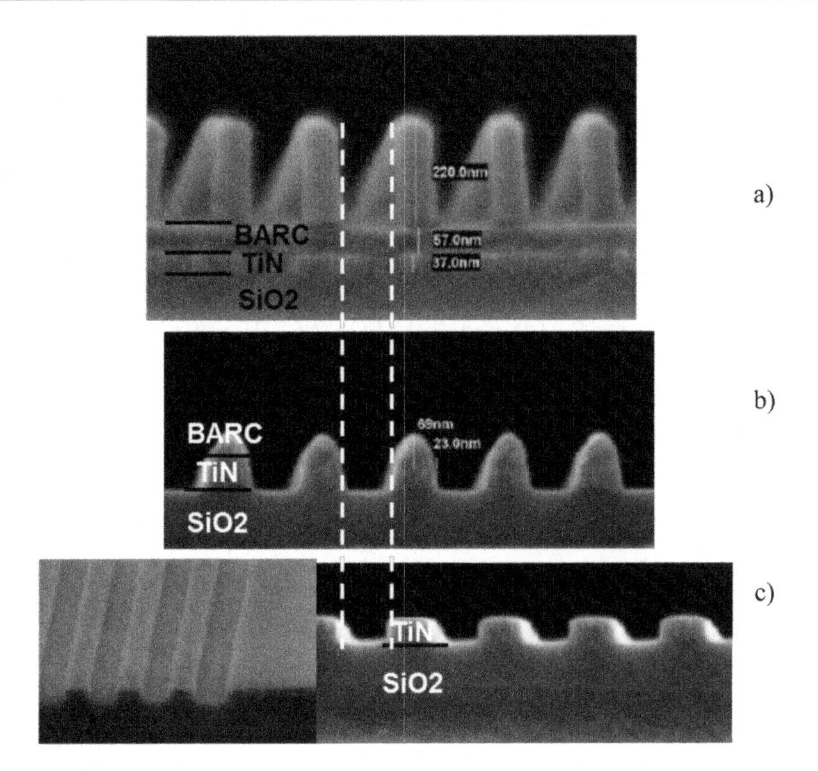

Figure 3.2. *Sequence of metal hard mask opening in a transformer-coupled plasma (TCP) etch tool: a) lithography, b) BARC and TiN opening followed by photoresist and BARC removal in a microwave plasma c)*

The choice of the chemistry for the TiN etching is very important because it directly controls the dimension of the line. The line dimension control can be achieved either during BARC opening or TiN etching by playing on plasma parameters like the radio frequency (RF) bias voltage (i.e. playing on ion energy) or process time (impacting the over-etch), as shown in Figures 3.3(a) and (b), respectively.

After TiN etching, the remaining PR and BARC are stripped away using H_2O_2 chemistry followed by O_2/N_2 chemistry using remote (downstream) plasmas (Figure 3.2(c)).

Following the PR and BARC stripping, a wet cleaning step is also performed in order to insure a complete removal of polymer residues. After the metal hard mask opening, the via patterning is then performed on the shallow hard mask topography, which is further planarized with the use of an organic BARC (Figure 3.1(c)). Then, the subsequent via etching, line etching and etch stop layer (ESL) opening (silicon-carbide-based film like, for instance, SiCN or SiC) are performed in one single etching process using fluorocarbon-based chemistry (Figures 3.1(d) and (e)) [POS 06]. If not mentioned, the etching is performed in a dual frequency capacitively coupled plasma (DFCCP) etch tool.

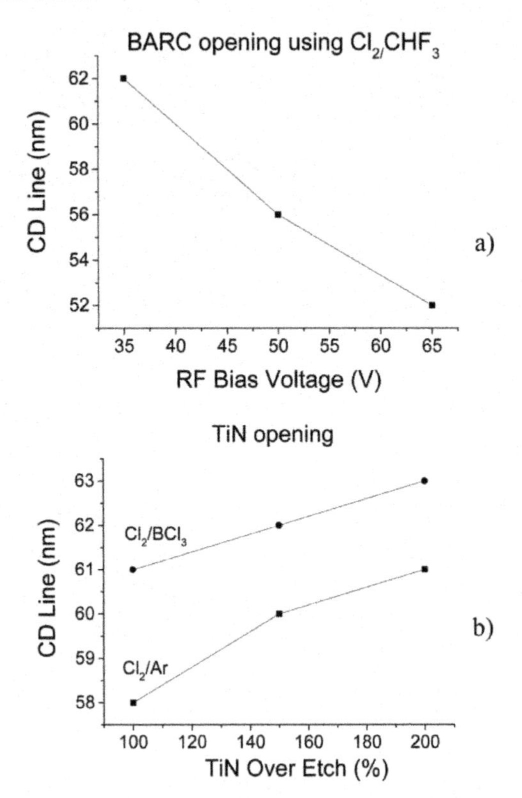

Figure 3.3. *Plasma parameter impact on critical dimension control of the line during either BARC a) or TiN b) opening*

3.1.2. *Via and line etching*

To ensure process stability and reproducibility during dielectric film etching with the metallic hard mask strategy, some hardware improvement must be done. Indeed, with this integration scheme, a drift of the line etching processes (etch rate, uniformity) and lower mean time between clean (MTBC) (100 h compared to 500 h with the use of an organic mask) is observed when using a non-heated top electrode, as shown in Figure 3.4(a). This is attributed to the accumulation of Ti-containing compounds on the reactor walls [CHE 07]. These compounds deposited on the top electrode lead to micromasking of the silicon electrode (also called "black silicon") which *in fine* degrades the chamber defectivity after several RF hours of etching processes. The solution to improving this chamber defectivity while using a metallic hard mask approach is to heat up the top electrode (above 100°C) to limit Ti species redeposition and thus silicon micromasking (Figure 3.4(b)). This allows the extension of the MTBC to more than 500 h. Today, for such integration schemes, the top electrode is heated at 120°C.

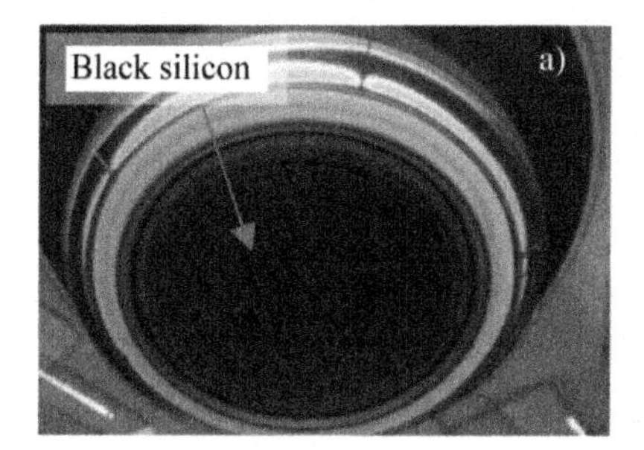

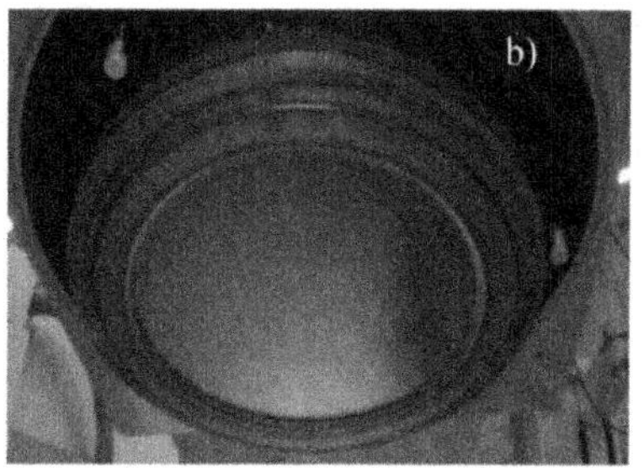

Figure 3.4. *Observation of a) non-heated and b) heated (>100°C) top electrodes after more than 500 RF hours processes of dielectric etching using a metallic hard mask strategy*

3.1.2.1. *Via etching*

The via (BARC and porous SiOCH films) is etched using fluorocarbon-based chemistry. For this step, the choice of the chemistry and plasma operating conditions for BARC opening is very important since it directly controls the dimension of the via transferred into the dielectric film. Indeed, for instance, an increase in the substrate temperature from 20 to 60°C leads to important CD increase after BARC opening using CF_4-based chemistry (Figure 3.5).

For the TFMHM integration, vias and lines are etched in one single etching process. For throughput requirements, it is mandatory to etch vias and lines at the same substrate temperature. The line etching temperature has to be above 50°C. Therefore, the partial via etching at such temperatures is very challenging as described below because of CD control (Figure 3.5).

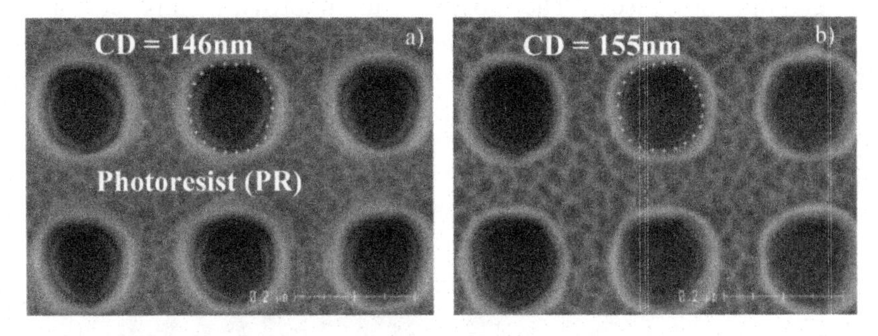

Figure 3.5. *Impact of the substrate temperature 20°C a) or 60°C b) during BARC opening using CF_4-based chemistry*

The solution to etching the vias at this temperature is to use highly polymerizing gas allowing a better CD control during the BARC opening step. Figure 3.6(a) shows a partial via profile obtained after BARC and porous SiOCH film etching using C_4F_6-based chemistry with a substrate temperature set at 60°C. The etch mechanisms of porous SiOCH film etching during partial via etching are similar to those presented in Chapter 2. Therefore, the use of a high polymerizing chemistry presents the benefit of having a better CD control at high substrate temperature but also presents the advantage of limiting the porous SiOCH film modification (see Chapter 2).

The remaining resist (PR) is then removed, as illustrated in Figure 3.6(b), using O_2-based chemistry at low pressure, which minimizes the low-k film damage (as described in Chapter 2). But contrary to the organic hard mask approach (presented later), the TFMHM integration provides a greater compatibility with porous dielectric materials since the modified porous SiOCH film, induced by the PR removal step, is removed during line etching.

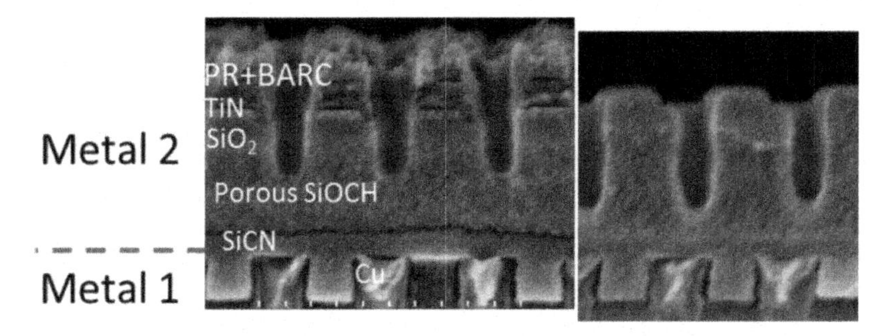

Figure 3.6. *Partial via etching in a DFCCP etch tool using a high polymerizing-based chemistry (left-hand picture) followed by a removal of the remaining resist using an oxygen-based chemistry (right-hand picture)*

3.1.2.2. *Line etching*

The porous SiOCH etching is performed using fluorocarbon-based chemistry. During this line etching process, the via is also etched to stop on the ESL.

Using a metal hard mask, the porous SiOCH etch mechanisms are similar to those presented in Chapter 2. During porous SiOCH line etching in fluorocarbon-based plasma, a high SiOCH:TiN selectivity (>100) is achieved since the main Ti-etched by-products (TiF_4) have relatively low volatility (boiling point at atmospheric pressure 284°C) as compared to SiF_4 (boiling point at atmospheric pressure –86°C) [DAR 06]. But the drawback of such low-volatile product formation is the risk of micromasking formation during dielectric etching. This micromasking formation is represented in Figure 3.7, after porous SiOCH (spin on, 50% void) etching in CF_4/Ar chemistry performed in a magnetic-enhanced reactive ion etcher (MERIE) with a substrate temperature set at 40°C. This behavior is assigned to Ti species redeposition on sidewalls and the bottom of trenches. In our experimental conditions, 12% titanium nitride (measured by X-ray photoelectron spectroscopy (XPS)) has been detected at the bottom of trenches [POS 05b].

Figure 3.7. *Illustration of porous SiOCH (spin on, 50% void) etching in a metallic hard mask environment with a CF_4/Ar etching chemistry performed in MERIE and a substrate temperature lower a) or higher b) than 50°C, respectively*

The solution to reducing this micromasking formation is to increase the substrate temperature above 50°C, as shown in Figure 3.7, to obtain higher volatility TiF_x compounds [DAR 08b]. In these conditions, Figure 3.7(b) shows lower micromasking formation, correlated with XPS analyses detecting only 5% of titanium at the bottom of trenches. Furthermore, the use of high polymerizing chemistry is also required to limit bottom-line roughness. The role of such chemistry is to reduce TiN hard mask etching. This solution is also in good agreement with the previous results presented in Chapter 2.

We previously showed that porosity introduction into dielectric film can lead to roughness formation during etching on blanket films. The present results are clearly evidence that the roughness formation during line etching is also strongly dependent on the plasma operating conditions with the metallic hard mask approach requiring the use of a substrate temperature above 50°C.

The first challenge for line etching is the high selectivity required between low-*k* and the ESL (SiCN). Indeed, low selectivity between both films can lead to copper sputtering or notch formation in the misalignment via (Figure 3.8(a)) during porous SiOCH (BD2x™) line etching performed in DFFCP using $C_4F_8/O_2/N_2/Ar$ chemistry. The use of highly polymerizing chemistry is mandatory during line etching to increase the selectivity between porous SiOCH and SiCN [POS 03]. Furthermore, as previously described, the use of highly polymerizing chemistry also presents the advantage of limiting porous SiOCH film damage.

Whatever chemistry and plasma etch conditions used, it is not possible to obtain a selectivity to the underlayer bigger than 4. This imposes adapting the TFMHM integration scheme. The partial via approach described in Figure 3.8(b) consists of etching only a part of the via during the via etching step. The remaining part of the via is etched during the line etching. Thus, the exposure time of the ESL is much shorter, relaxing the constraint on the etch selectivity and avoiding the notch formation.

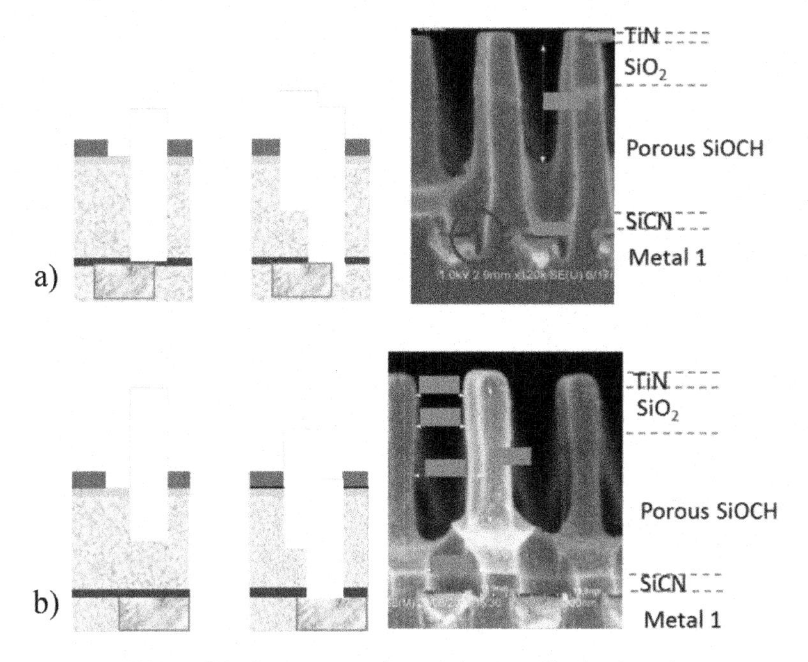

Figure 3.8. *Comparison of trench first metallic hard mask integration after line etching followed by etch stop layer opening for either full via a) or partial via approach b)*

Therefore, porous SiOCH film etching using a metallic hard mask is very challenging in terms of chamber defectivity and profile control. But using a heated top electrode, a substrate temperature above 50°C and highly polymerizing chemistry can overcome these issues. In this manner, acceptable profile and CD control can be achieved as shown in Figure 3.8(b) (porous SiOCH film etching using $C_4F_8/O_2/N_2/Ar$).

Thus, the TFMHM is an integration of choice for porous SiOCH film introduction. However, etching processes involved during the porous SiOCH line etching combined with the metal hard mask strategy generate other serious issues such as residue growth on the top and the sidewall of trenches after air exposure but also wiggling phenomena of the patterned stack. These issues will now be discussed.

3.1.3. *Residue formation*

One of the most critical issues with the use of a metallic hard mask for dielectric films integration is the formation of residues. Indeed, once dielectric films (SiO_2, SiOCH or porous SiOCH) are etched in fluorocarbon-based plasmas using a metallic hard mask, followed by air exposure, residues grow after a few hours of air exposure [POS 10].

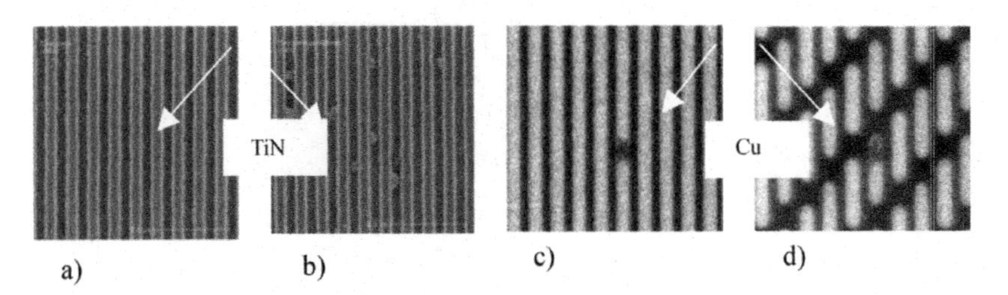

Figure 3.9. *Top CD SEM images of porous SiOCH (BD2x^{TM}) etching using $C_4F_8/N_2/Ar/O_2$ performed in DFCCP a) view showing residue growth after 24 h air exposure b) with or without wet cleans and their impact on defect generated in the line c) and via d) after CMP*

Those residues (Figure 3.9(b)) growing as a function of air exposure time are not removed after conventional wet cleaning (basic solution + 0.1% hydrofluoric acid (HF)). They prevent a good conformity during metal barrier and copper deposition generating via or line opens (Figures 3.9(c) and (d), respectively) after chemical mechanical polishing (CMP), strongly degrading the electrical yield (Figure 3.10).

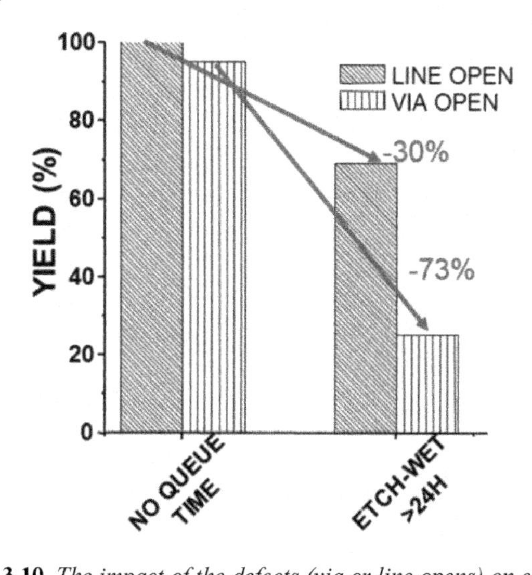

Figure 3.10. *The impact of the defects (via or line opens) on electrical performance has been shown by monitoring the yield for minimum pitch 70/70 nm as a function of the waiting time, between the etch and wet steps for a C045 metal 2/via 1 (M2/V1) interconnect level*

With the continuous downscaling in dimension of integrated circuits, this residue formation became the main defectivity and yielded a killer factor from the 45 nm technological node.

3.1.3.1. *Mechanism of residue formation*

In order to have a better understanding of these residue formation mechanisms, further analyses were performed on titanium nitride blanket wafers deposited by physical vapor deposition. The pristine film shows a clean surface (Figure 3.11(a)). The analyzed surface of the pristine film, determined by XPS, is mainly composed of titanium (23%), nitrogen (22 %) carbon (21%) and oxygen (34%). After exposition to $C_4F_8/N_2/Ar/O_2$ etch chemistry performed in DFCCP, important residues are observed (Figure 3.11(b)). In this case, an important fluorocarbon layer is formed on the top of the TiN.

We demonstrated that the main factors responsible for the defect formation are attributed to fluorine species present on the metal hard mask and low-*k* surfaces after plasma exposure and air moisture. Indeed, fluorine species react with air moisture from the atmosphere leading to HF that forms metallic residue with titanium [POS 10].

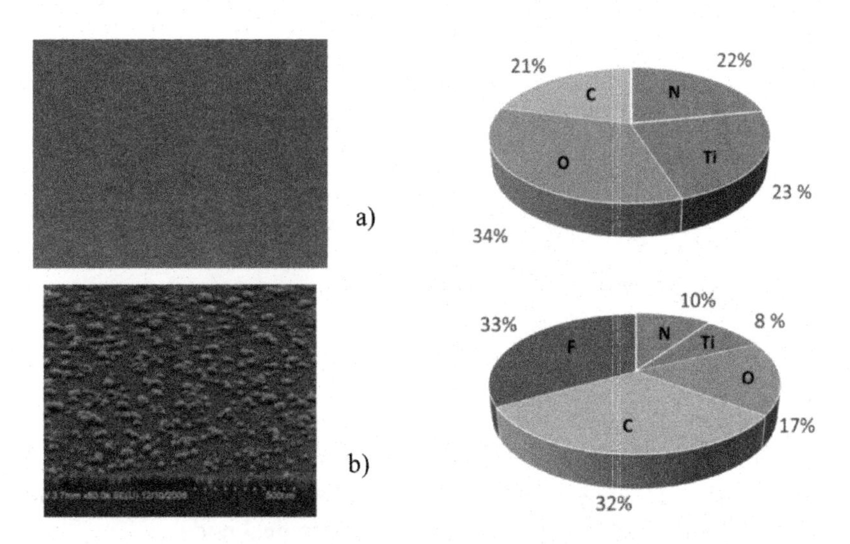

Figure 3.11. *Top CD SEM observation (on the left hand) and associated XPS composition (on the right hand) of titanium nitride film before a) and after exposure to $C_4F_8/N_2/Ar/O_2$ etch chemistry b) performed in DFCCP*

This mechanism is confirmed by residue observed after exposing the titanium nitride pristine film to HF (1%) vapor during 30 sec residue (Figure 3.12(a)). A similar result is observed when the titanium nitride is exposed to HCl (Figure 3.12(b)).

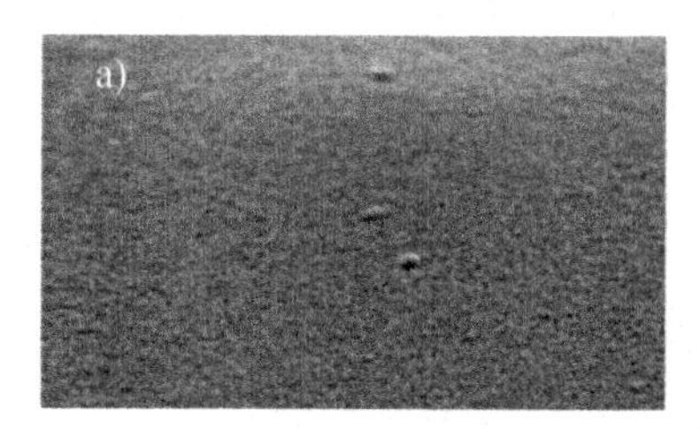

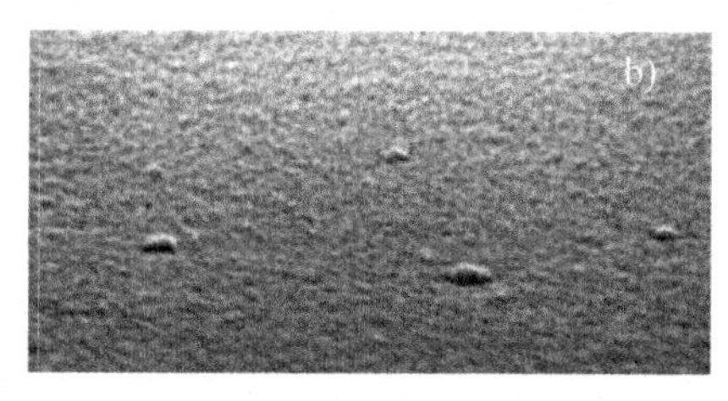

Figure 3.12. *Top CD-SEM view of the pristine TiN exposed to 1% vapor HF a) and HCl b) for 30 sec*

More generally, the mechanism responsible for the residue formation on a metallic surface is the following:

$$\text{Metal} + \text{Acid} \rightarrow \text{Residue} \quad [3.1]$$

3.1.3.2. *Solution to limit residue formation*

A potential solution to avoiding their formation is to prevent reaction between fluorine and air moisture. This can be done by either removing the fluorine or avoiding the reaction with the air moisture.

In situ post-etch plasma treatments such as oxidizing or reducing chemistries have been proposed to limit the residue formation [POS 11, KAB 13].

The efficiency of these post-etch treatments is shown in Figure 3.13. After exposition of titanium nitride to C_4F_8 /N_2/Ar/O_2 etch chemistry followed by *in situ* O_2 or H_2 post-etch treatment performed in DFCCP etch tool, a clean surface is observed in both cases.

In this case, oxidizing and reducing chemistries limit the residue formation by partially removing fluorine from the TiN surface to 8% and 13%, respectively (compared to 33% fluorine measured after etching). These results show that such post-etch treatments are efficient for limiting the residue formation but their efficiency is still dependent on the fluorine concentration present at the surface of the wafer. These post-etch treatments delay the formation of residues on metal hard mask under atmosphere due to a decrease in fluorine concentration, but do not stop the reaction.

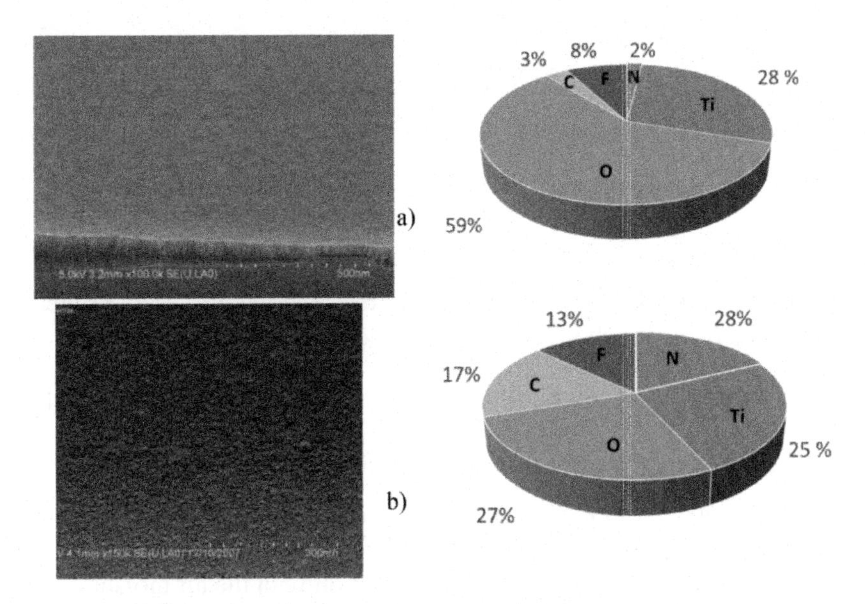

Figure 3.13. *Top CD SEM observation (on the left hand) and associated XPS composition (on the right hand) of titanium nitride film after exposure to $C_4F_8/N_2/Ar/O_2$ etch and in situ oxidizing a) or reducing b) post-etch treatment performed in DFCCP*

Another way to limit the reaction between fluorine and air moisture is to form a passivation layer on top of the titanium nitride to prevent the reaction with air moisture. This can be achieved by using methane-based post-etch treatment. In this case, the carbon-rich passivation layer delays the reaction between fluorine and air moisture. The benefit of such post-etch treatment is depicted in Figure 3.14. Despite important fluorine concentration being detected on top of titanium nitride, a clean surface is observed.

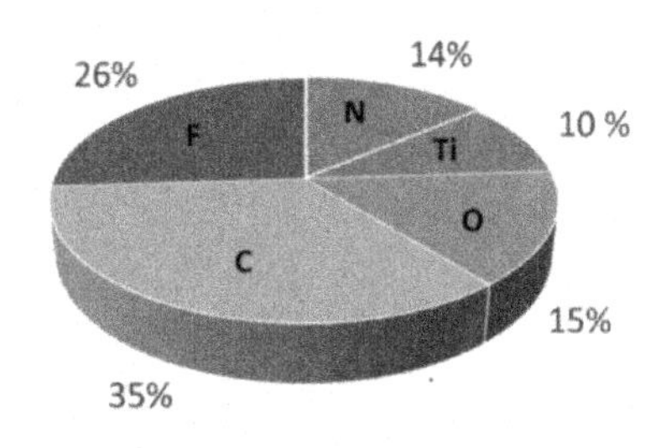

Figure 3.14. *Top CD SEM observation (on the left-hand side) and associated XPS composition (on the right-hand side) of titanium nitride film after exposure to $C_4F_8/N_2/Ar/O_2$ etch chemistry followed by in situ methane-based post-etch treatment performed in DFCCP*

Under our experimental conditions, the methane-based post-etch treatment presents higher performance than any other chemistries investigated since its efficiency of limiting the residue growth and being independent of Flurocarbon (FC) etching process conditions [POS 11]. The efficiency of the different post-etch treatment described previously has been confirmed on patterned wafers (Figure 3.15).

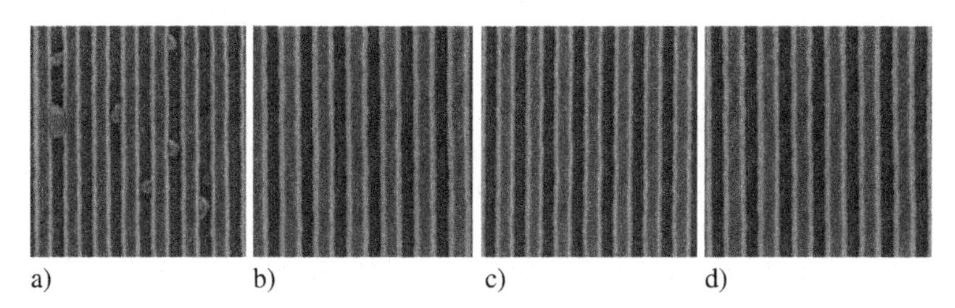

Figure 3.15. *a) Top CD SEM view of metal 1 level after porous SiOCH (BD2x^{TM}) $C_4F_8/N_2/Ar/O_2$ etching a) followed by in situ H_2 b), O_2 c) or CH_4 d) post-etch treatment performed in DFCCP plus 24 h air exposure*

3.1.3.3. *Post-etch treatment compatibility with porous SiOCH film*

As described in Chapter 2, the risk with such chemistries is to induce the degradation of important porous SiOCH film properties. Figure 3.16 shows the impact of the different post-etch treatments on porous SiOCH (BD2x^{TM}) integrity using the decoration method.

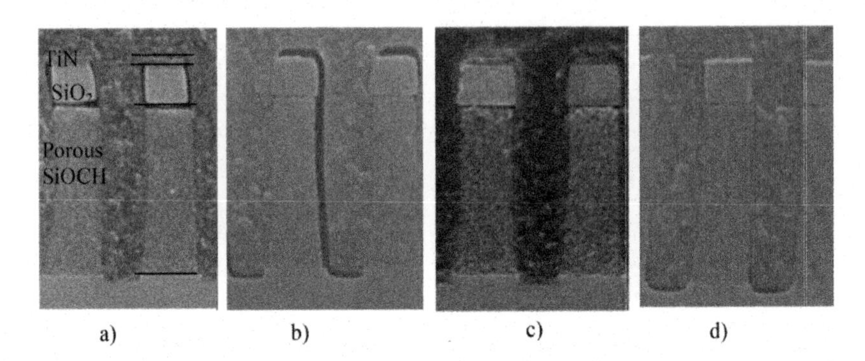

Figure 3.16. *Sidewall film modification of porous SiOCH (PECVD, 25% void)) after etching $C_4F_8/N_2/Ar/O_2$ a) as a function of the post-etch treatment added O_2 b), H_2 c) and CH_4 d). The sidewall modification is determined using the decoration method*

Oxidizing post-etch treatment is prohibited for porous SiOCH film since important film degradation of the dielectric film is observed (Figure 3.16(b)). Hydrogen- or methane-based chemistry presents a better compatibility with the dielectric film (Figures 3.16(c) and (d), respectively). This compatibility has been confirmed by complementary electrical and reliability tests.

Figure 3.17(a) shows copper line resistance versus intraline capacitance distribution with or without methane-based post-etch treatment addition. The test structure is a serpentine comb structure with a total length of 14,000 µm and spacing between lines of 70 nm. The resistance–capacitance (RC) distribution is similar with or without post-etch treatment addition, indicating that such a treatment has no impact on electrical performances. The post-etch treatment does not modify the capacitance with respect to the process of record (POR), indicating that the relative permittivity of the dielectric line remains constant. A similar trend is observed after hydrogen post-etch treatment addition to the etch POR.

Figure 3.17(b) shows that methane-based post-etch treatment has no impact on electromigration (EM) reliability performed for a metal X (Mx) level integration (a similar result for M1). In our experiments, EM tests have been performed on a metal 2/via 1 structure at 300°C and 0.2 mA. Time-dependent dielectric breadkdown (TDDB) tests representing cumulative time to failure distribution at 28 V and 125°C for a metal 1 structure show that methane-based post-etch treatment addition to the POR enhances the TDDB (Figure 3.17(c)).

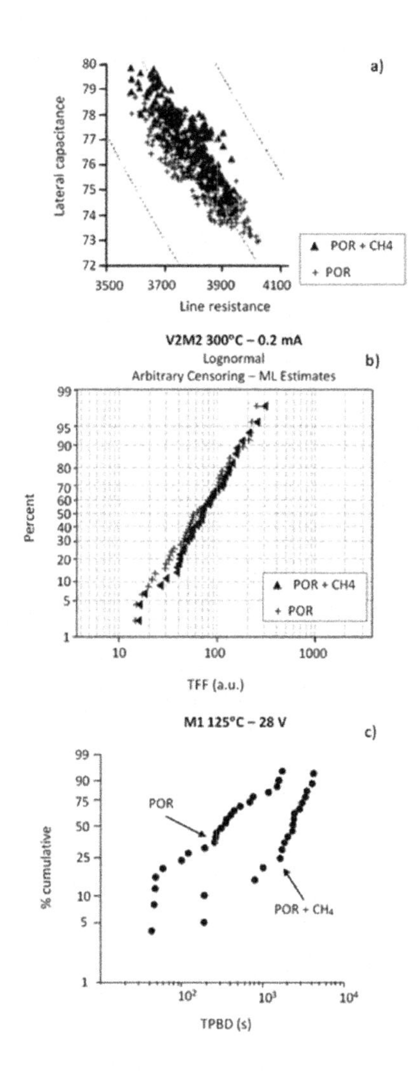

Figure 3.17. *Impact of methane post-etch treatment addition to a process of reference on interline capacitance versus line resistance for different chemistries: line width = 70 nm and interline space = 70 nm a), electromigration b) and time to breakdown c)*

3.1.3.4. *Synthesis*

Oxidizing or reducing *in situ* post-etch treatments limits the residue formation by partially removing fluorine from the TiN surface, while their efficiency depends on the fluorine concentration present on the titanium nitride. Despite its efficiency in limiting the residue formation, the oxidizing chemistry is prohibited since it induces important porous low-*k* film damage. Using hydrogen chemistry, no (or lower) porous film damage has been observed confirming the results presented in Chapter 2.

Using a methane-based chemistry, the formation of a carbon-based passivation layer on the top of the titanium nitride prevents the reaction with air moisture leading to a residue-free surface even after one day air exposure. In these conditions, this post-etch treatment presents higher potential than oxidizing or reducing chemistries since its efficiency to limit residue growth is independent of the fluorocarbon etching process conditions. In our experimental conditions, such a post-etch treatment showed its compatibility with porous SiOCH film integration.

Finally, based on these observations, a combination of two post-etch treatments, such as a reducing chemistry (to remove the fluorine at the surface) followed by a methane-based chemistry (to form a carbon passivation to prevent reaction with air moisture), appears to be one of the most promising solutions.

3.1.4. *Line wiggling*

With the constant downscaling in dimension, etching process steps are facing new issues. Indeed, patterning of sub-50 nm porous SiOCH trenches using a metallic hard mask approach leads to a line undulation phenomenon (Figure 3.18), which is also called line wiggling [DAR 07, DUC 14].

Darnon *et al.* [DAR 07] demonstrated in their experimental conditions (high compressive stress –2,400 MPa, with dielectric materials having low elastic properties (= 0.9 GPa)) that wiggling of sub-100 nm porous dielectric lines can occur with a metallic hard mask strategy. They assigned this wiggling phenomenon to the low mechanical property of the dielectric material coupled to the high compressive stress of the metallic hard mask. This indicates that this mechanical limitation could become a real issue for advanced interconnect technology nodes.

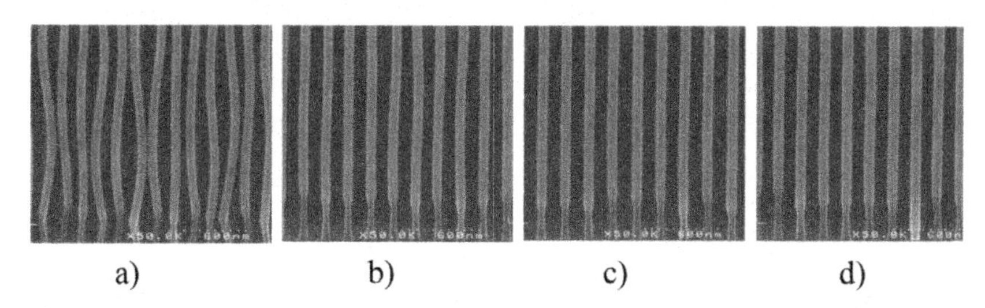

a) b) c) d)

Figure 3.18. *Profile evolution after $C_4F_8/N_2/Ar/O_2$ line etching performed in DFCCP of 290 nm porous SiOCH (BD2x^{TM}) with residual titanium nitride hard mask stress of –2,250 MPa for different line dimensions: a) 35, b) 45, c) 50 and d) 55 nm [DUC 14]*

To explain the origin of the line undulations, mechanical simulations were performed using the commercial mechanical finite-element analysis code, ANSYS™. The buckling analysis is for bifurcation buckling using a linearized model of elastic stability. The simulated structure used for buckling analysis consists of a 1-µm long line with the stack presented in Figure 3.19.

The matrix of stiffness [S] and the matrix of stress stiffness are calculated from this structure based on iterative calculation of the equilibrium state. From buckling analysis, a coefficient is extracted by which the matrix of stress stiffness [S] has to be multiplied to reach the first buckling Eigen mode. This so-called buckling coefficient is critical when it lies under unity. More details about the experimental protocol can be found in previous studies [DAR 07, DUC 14].

Due to these simulations, the impact of parameters such as TiN film stress and thickness, line width and height and materials' Young modulus, potentially responsible for dielectric lines undulations, has been investigated using finite-element mechanical modeling.

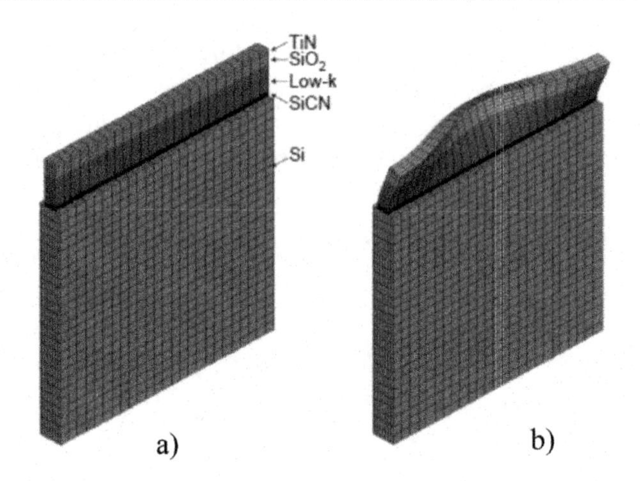

Figure 3.19. *Example of simulated stack before a) and after deformation calculation b) for metal 1*

3.1.4.1. *TiN stress and thickness effect on the buckling parameter*

With the model just described, the effect of the residual stress and thickness impact on the buckling of dielectric lines can be simulated as a function of the dielectric line width (see Figure 3.20).

Each curve shown in this graph corresponds to a buckling coefficient of 1 and separates the line width/residual stress plan into two areas. For each parameter, the left part of the plan delimitated by the curve corresponds to a buckling coefficient lower than 1 (wiggling), while the right part corresponds to a buckling coefficient higher than 1 (no wiggling).

As shown in Figure 3.20, for a 45-nm-thick TiN mask having highly compressive residual stress (–2,200 MPa), no wiggling is expected down to below 35-nm-wide lines. Reducing the TiN thickness down to 15 nm allows relaxing the wiggling phenomenon down to 20-nm-wide features. With lower TiN compressive stress (–750 MPa) and similar thickness (45 nm), no wiggling is expected for line widths down to less than 20 nm.

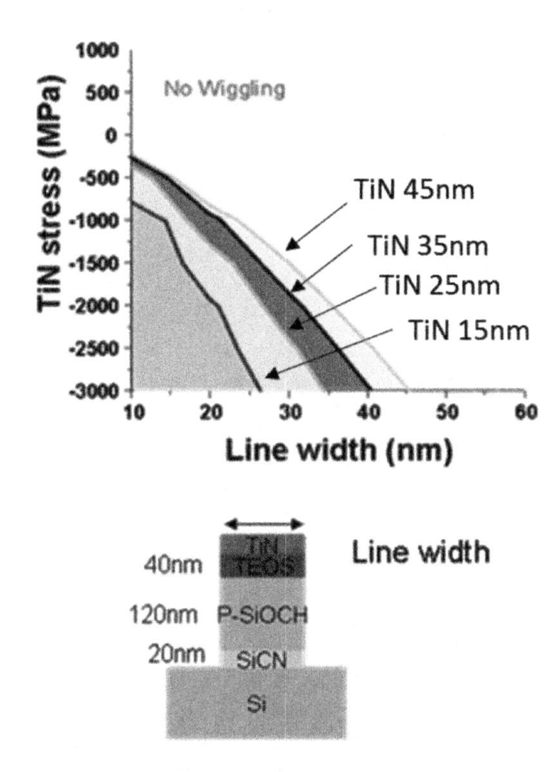

Figure 3.20. *Simulation of the wiggling phenomenon as a function of the line width, TiN residual stress and thickness from a C032 stack*

These simulations clearly evidence that TiN residual stress and its thickness play a key role on the line undulation.

3.1.4.2. *Porous SiOCH film thickness effect on buckling parameter*

The effect of the porous SiOCH thickness on the wiggling phenomenon was also simulated by varying the thickness between 20 and 300 nm for a highly compressive TiN stress (–2,200 MPa). Figure 3.21 shows that when the porous SiOCH film thickness increases, the line width for which the wiggling phenomenon is observed increases as well.

This simulation shows that the wiggling also depends on the porous SiOCH thickness.

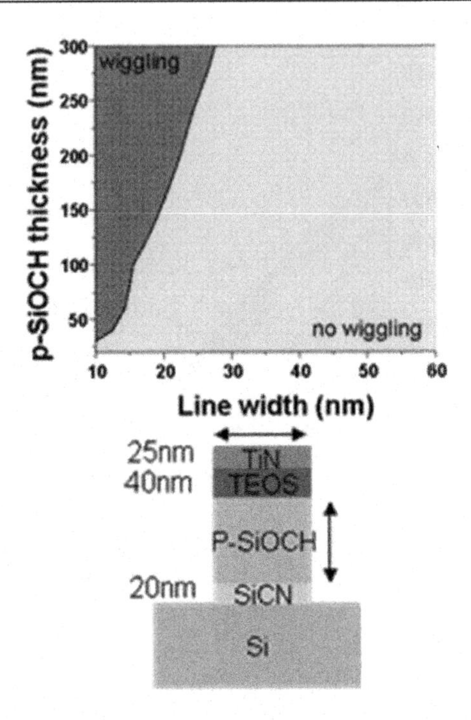

Figure 3.21. *Simulation of wiggling as a function of line width and porous SiOCH (p-SiOCH) thickness with a TiN stress (–2,200 MPa) and thickness (25 nm)*

3.1.4.3. *TiN and porous SiOCH Young modulus effects on buckling parameter*

The effect of TiN and porous SiOCH Young moduli on the line wiggling is represented in Figures 3.22(a) and (b), respectively. These simulations suggest that a TiN or porous SiOCH Young modulus variation of 200 or 10 GPa, respectively, has a low impact on the wiggling line formation since it is only allowing the relaxation of the wiggling formation for line dimension of approximately 5 nm.

Therefore, TiN and porous SiOCH Young moduli have an impact on wiggling formation but in second order compared to the impact of the TiN stress and thickness.

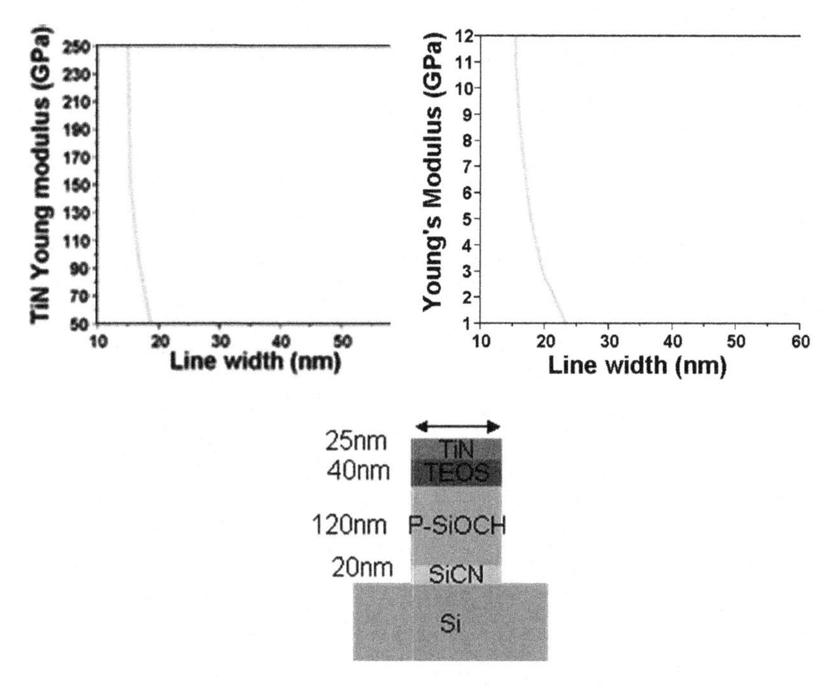

Figure 3.22. *TiN a) and porous SiOCH b) Young modulus impact on wiggling formation as a function of the line width, with a TiN stress and thickness set at –1,000 MPa and 25 nm, respectively*

3.1.4.4. *Summary*

The TiN (stress and thickness) combined with the porous SiOCH film thickness has a strong impact on the wiggling formation. But the effect of the porous SiOCH film thickness on line undulation is mitigated for advanced technology by the continuous decrease in porous SiOCH thickness. Therefore, one solution to continuing porous SiOCH material integration with metallic hard mask for advanced interconnect technology node is to develop lower TiN residual stress and thickness (Figure 3.23).

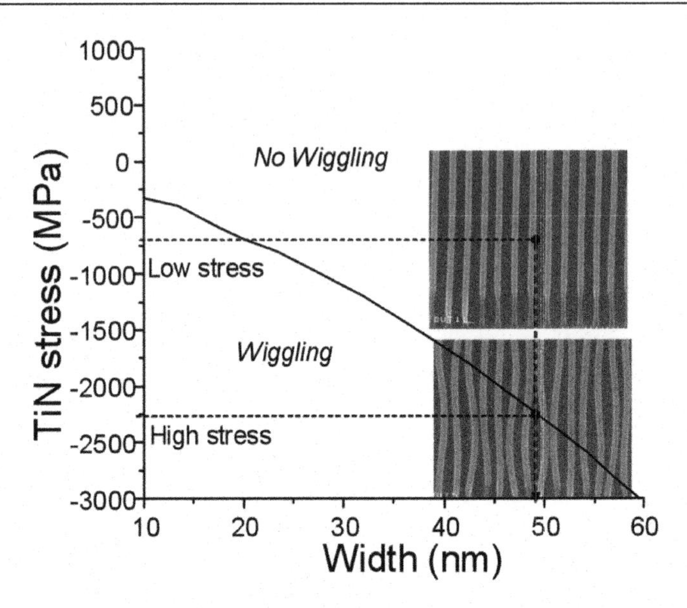

Figure 3.23. *Comparison of the line wiggling formation between simulation prediction and experimental results as a function of the line dimension for a stack made up of 20 nm SiCN/290 nm porous SiOCH (BD2x^{TM})/40 nm SiO_2 with high and low TiN (35 nm) residual stress*

3.2. Porous SiOCH integration using the via first approach

Contrary to the TFMHM scheme, the via is etched before the line with the "via first approach". As presented in Figure 3.24, the first lithography step is used to define the via. Then, the mask is removed and a planarizing layer (PL) is deposited to form a plug in the via and a mask for the line. The PL is capped with dielectric and the second lithography step is used to define the line in a so-called trilayer mask. The capping and the PL are etched to form the mask while leaving the plug at the bottom of the via. Then, the trench is etched inside the low-*k* and the PL is removed. Finally, the dielectric barrier is open.

Contrary to the TFMHM scheme, the low-*k* dielectric is exposed to the plasma during the mask removal process. For this reason, the via first integration scheme (standardly used) has been progressively replaced by the TFMHM over the years. We will briefly describe here the specificities and challenges of the via first integration scheme.

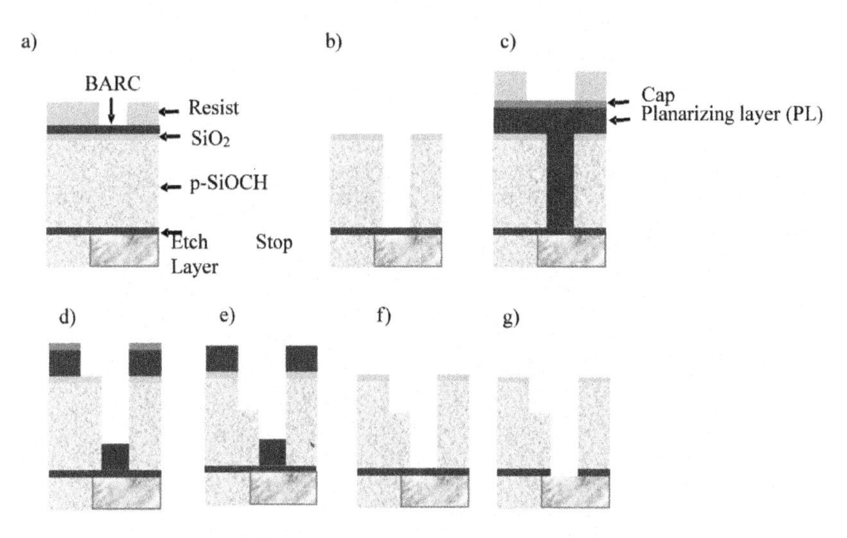

Figure 3.24. *Schematic of the process flow for the via first hard mask integration scheme*

3.2.1. *Via etching*

The etch process and challenges for the via etching are identical to the process used with the trench first metal hard mask (see Figure 3.1). Indeed, in both cases, a PR mask is used to define the pattern. In addition to the requirements shown with the TFMHM, the via first integration scheme needs to etch the via down to the dielectric barrier while being selective to the dielectric barrier. This leads to vias with larger aspect ratios. The via etching is usually performed in two steps. The first step, called "main etch", is used to etch the via with a process optimized for the via profile. The second step, called "over-etch", finishes the via etching with a process highly selective to the dielectric barrier. Contrary to the trench first metal hard mask integration scheme, there is no requirement on the wafer temperature during the via etching since the line and via etching are not performed in the same sequence. In addition, we will see that the line etching can be performed at room temperature with an organic mask.

Figure 3.25 presents a cross-section scanning electron microscopy (SEM) of two vias etched in a porous low-*k*. In this case, the dielectric is deposited by spin coating and has a *k* value of 2.2.

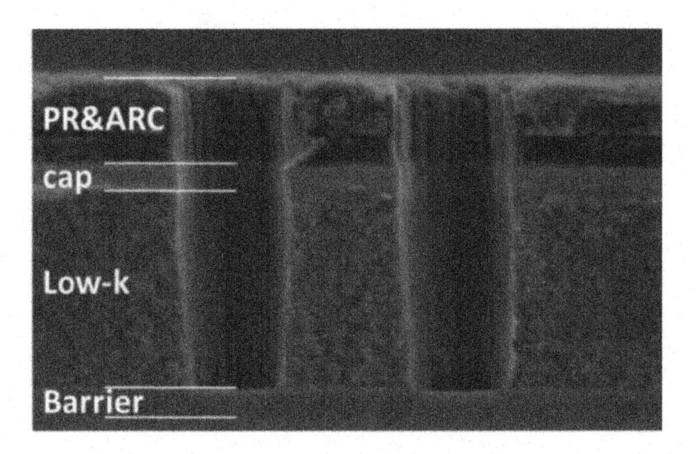

Figure 3.25. *Cross-section SEM of vias etched in a porous low-k dielectric. PR and ARC stands for photoresist and anti-reflective coating, respectively*

3.2.2. *Trilayer mask etching*

The trilayer mask strategy consists of using planarizing, capping and PR layers. Such trilayer masks are also used for other etching applications in the semiconductor industry when the PR is too thin and fragile to sustain the plasma processes [ABD 08]. In this case, the patterns are defined in the thin and fragile PR by lithography, and plasma etching steps are used to transfer the patterns from the PR to the capping layer and then from the capping layer to the underlayer (in our case, the PL). For back-end-of-line applications, all the steps are performed in the same chamber to optimize the process flow. As a result, trilayer mask opening, line etching and mask removal are all performed with capacitively coupled plasmas.

The first step of the mask opening is the capping layer etching. The capping layer is usually made up of few tens of nanometers of SiO_2 or SiOCH. The etching is performed using a fluorocarbon-based plasma. During this step, it is possible to play on the capping layer profile to slightly increase (e.g. by adding oxygen in the plasma) or reduce the trench width (e.g. by adding a polymerizing gas in the plasma) to correct the trench dimension defined by lithography, as shown in Figure 3.26. This is particularly useful when feed-forward-advanced process control is used to correct lithography process instabilities.

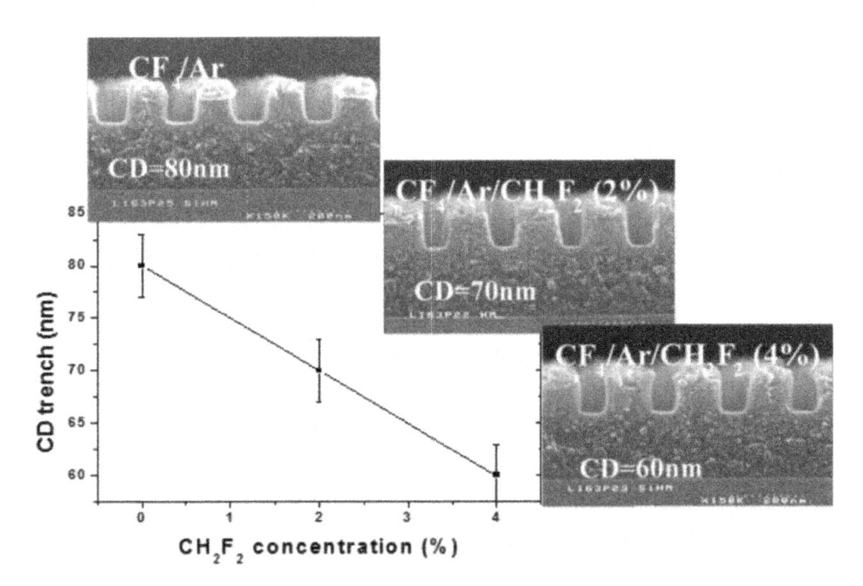

Figure 3.26. *Evolution of the trench width as a function of the CH_2F_2 gas concentration during the cap opening process. CD stands for critical dimensions. The associated cross-section SEM pictures are also reported*

The second step consists of etching the PL using the capping layer as a mask. PLs are commonly polymers that can be etched in oxygen- or hydrogen-based plasmas [DAR 10]. During the first few seconds of the process, the PR is removed by the etching plasma. The challenge with the PL etching resides in controlling the mask profile and dimensions. In particular, oxygen-based plasmas are very reactive with polymers and lateral etching can be observed if there is no passivation layers at the pattern sidewalls to block the etching [DAR 10, FUA 01].

With hydrogen-based plasmas, the reactivity is reduced and heavy low-volatility etch by-products can also deposit at the sidewall and reduce the lateral etching [PON 94]. Figure 3.27 compares the pattern profile during PL etching in N_2/O_2 or NH_3 plasma [DAR 10]. The trench width increases during the etching in O_2/N_2 plasma because of mask faceting and lateral etching while straight profiles that tend to bow during the over-etch are obtained with the NH_3 plasma. The over-etch process must be well-controlled so that the mask is open in all trench areas but some PL remains as a plug at the bottom of the vias.

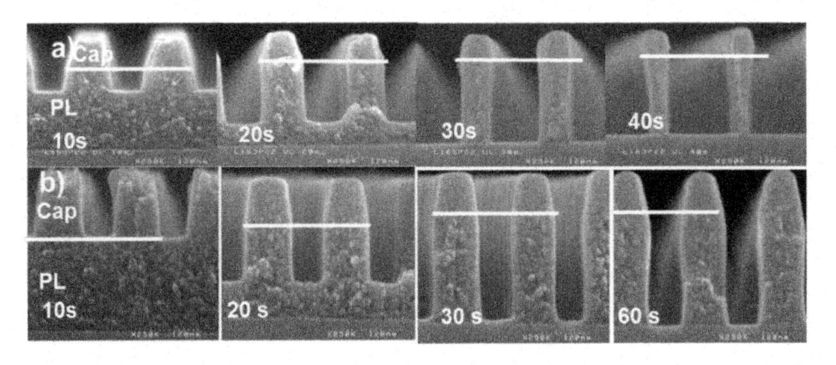

Figure 3.27. *Pattern profile evolution during the etching of the planarizing layer in a) N_2/O_2 plasma and b) NH_3 plasma*

3.2.3. *Line etching*

The mask defined by the PL is used to etch the dielectric layer, while the bottom of the via is protected by a PL plug. As presented in Chapter 2, fluorocarbon-based plasmas are standard recipes for low-*k* etching [DAR 10, POS 04]. As shown in Figure 3.28, the selectivity and the trench profile can be tuned by changing the polymerizing rate of the plasma. As explained in Chapter 2, polymerizing plasmas are preferred since they ensure a low surface roughness and minimize plasma-induced damage.

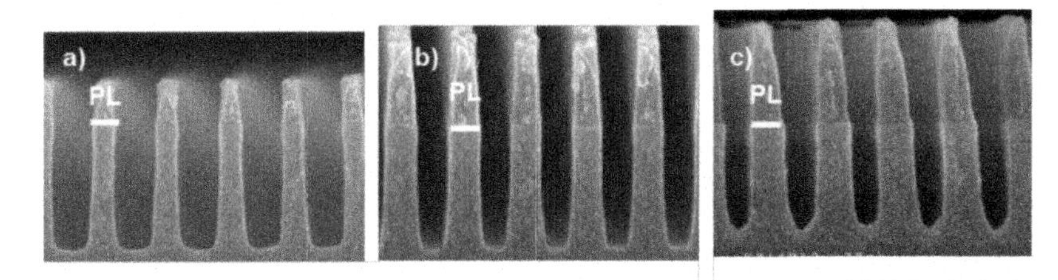

Figure 3.28. *Low-k trench profile after low-k etching in an MERIE plasma using a) non-polymerzing process (CF_4/Ar), b) mild polymerizing conditions (CF_4/Ar with 2% CH_2F_2) and c) highly polymerizing conditions (CF_4/Ar with 4% CH_2F_2) [DAR 10]*

3.2.4. *Mask removal*

After the line has been etched, the PL material must be removed from the bottom of the via to enable the barrier etching. The main challenge for the mask removal is to remove a carbon-based material without consuming the carbon contained inside the porous materials. Indeed, as explained in Chapter 2, the porous low-*k* dielectrics can be damaged during oxygen- and hydrogen-based plasmas that are used for organic materials removal. In addition to the plasma-induced damage, the fluorocarbon species deposited on the chamber walls are also consumed during the mask removal process, feeding the plasma with fluorine that can participate in the porous low-*k* damage and/or perturb the patterns profile. To overcome these issues, it is possible to perform the mask removal process in a separate chamber operating downstream hydrogen-based plasmas at high temperature that are not contaminated with fluorine and that are known to remove carbon-based masks without inducing large material modification [POS 07]. However, this approach increases the process flow complexity.

3.2.5. *Barrier opening*

The final step of the etching process is the barrier opening. As for the TFMHM integration scheme, the bottom of the trenches is also partially etched during the barrier etching step, and the pattern profile can be partially modified, in particular at the edges between the vias and the trench that can be facetted. The chamfer resulting shape can be beneficial for the metallization steps, provided that its extent is controlled.

Figure 3.29 presents the pattern profile in a via chain before trench etching (Figure 3.29(a)), and after trench etching (Figure 3.29(b)), mask removal (Figure 3.29(c)) and barrier opening (Figure 3.29(d)). The low-*k* dielectric is deposited by spin coating and has a dielectric constant of 2.2.

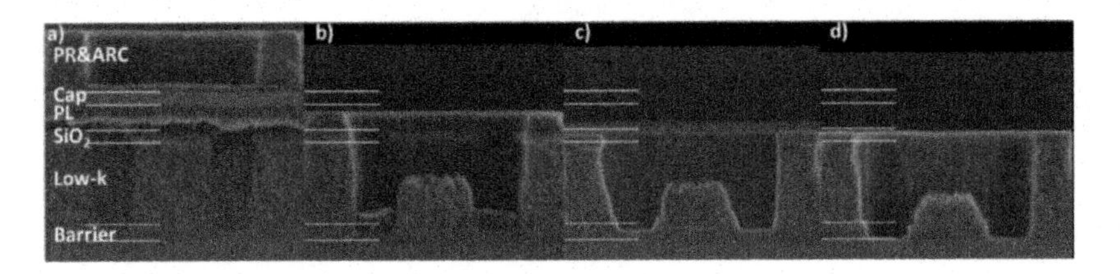

Figure 3.29. *Cross-section SEM pictures of a via chain structure a) before etching, b) after trilayer mask opening and trench etching, c) after mask removal and d) after barrier opening*

3.2.6. *Limitations of the via first integration scheme*

The main limitation associated with the via first integration scheme resides in the necessity to remove the PL after the trench patterning. During this step, the porous low-*k* dielectric is exposed to the plasma and may be damaged [POS 04]. This results in a degradation of the dielectric properties and of the circuit performance and reliability.

With the reduction of the pattern dimensions, an additional limitation related to the pattern stability occurs. The so-called pattern collapse or pattern flop over comes from the bending of tall and narrow structures (see Figure 3.30). When the aspect ratio of the structure (either the planarization layer only or the stack of low-*k* and PL) is large, small forces can lead to large bending that may overpass the elasticity limit of the structure. The small forces can originate from surface tensions, electrostatic forces and/or inertia forces. The theory of mechanics predicts that the deflection of an embedded beam scales with the cube of the beam aspect ratio and the inverse of the Young modulus (see Figure 3.30). Since the etch selectivity between the PL and the low-*k* is low (typically <10), a thick mask is needed, leading to a large aspect ratio that becomes larger when the line width is reduced. As a result, for the smallest dimensions, the use of the via first approach can lead to pattern collapse, which can, however, be reduced by using planarizing materials with a high Young modulus and/or a good etch resistance.

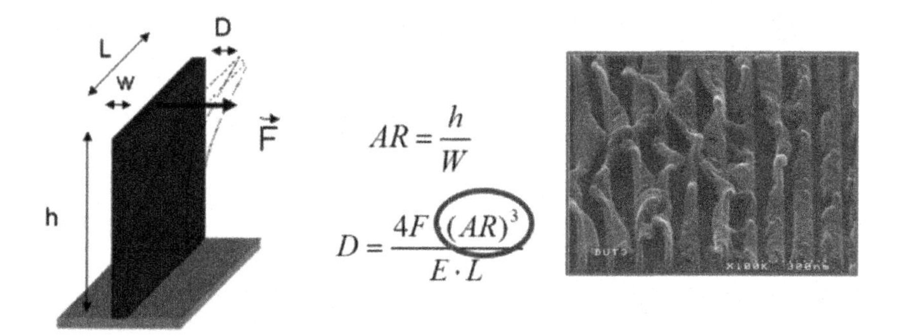

Figure 3.30. *Schematic view of an embedded beam and cross-section SEM pictures of high aspect ratio polymer lines that are partially collapsed*

3.3. Summary

In this section, we presented the two integrations schemes (TFMHM (Figure 3.1) and via first (Figure 3.24)) proposed for low-*k* film integration. Both integration schemes present advantages and drawbacks as summarized in Table 3.1.

With the introduction of porosity into the SiOCH materials at the 45 nm node, the metallic hard mask strategy appeared to be the most promising integration scheme (despite its drawbacks such as residue formation, line wiggling and problems that are well known and solved today) since no ashing steps are required after line etching.

	Trench first metallic hard mask	Via first
Mask selectivity	>100	>10
Barrier Selectivity	Required partial via	Barrier protection by plug
Defectivity	Residue/chamber wall	Low
Scaling limitation	Wiggling	Pattern collapse
Film damage	Etching	Etching and ashing
Process flow complexity	Low	High

Table 3.1. *Comparison of trench first metallic hard mask and via first approach for low-k film integration*

4

Interconnects for Tomorrow

The need to decrease the resistive-capacitive (RC) delay, dynamic power consumption and cross-talk noise in back-end-of-line requires the use of porous dielectrics as interconnect insulators. The major issue with porous film integration is their sensitivity to plasma processes that consume methyl groups and make the low-*k* hygroscopic. We have seen that both plasma etching and post-etch plasma treatments lead to film modification. The species diffusion through the pore network is responsible for the in-depth modification of porous low-*k*.

In the following chapter, we will discuss the impact of porosity increase on integration difficulties in dielectric film, required to continue device performance improvement. We will also give an overview of disruptive solutions to further reduce the interconnect insulator dielectric constant.

4.1. Consequence of porosity increase

As we have discussed earlier, a material's sensitivity to plasma processes is directly linked to the porosity that provides the diffusion path for plasma species. Increasing the porosity directly favors the diffusion of species, and amplifies plasma-induced damage. As a

---

Chapter written by Maxime DARNON and Nicolas POSSEME.

result, plasma-induced damage of porous materials becomes increasingly challenging as long as the porosity increases [MAE 13, WAN 84]. With current plasma processes, it is not possible to integrate porous low-*k* dielectrics without inducing damage.

The increase of porosity also brings other challenges for porous low-*k* integration. Indeed, as shown in Figure 4.1, the mechanical stability of porous materials decreases when the porosity increases, which leads to defectivity during the mechanical polishing or the packaging processes [IAC 04]. Adhesion is also low and cracks, or dicing and wafer bonding, form during the chemical mechanical polishing (CMP). A further increase of the porosity would also lead to soft materials for the integration unless the mechanical stability of the low-*k* matrix is improved.

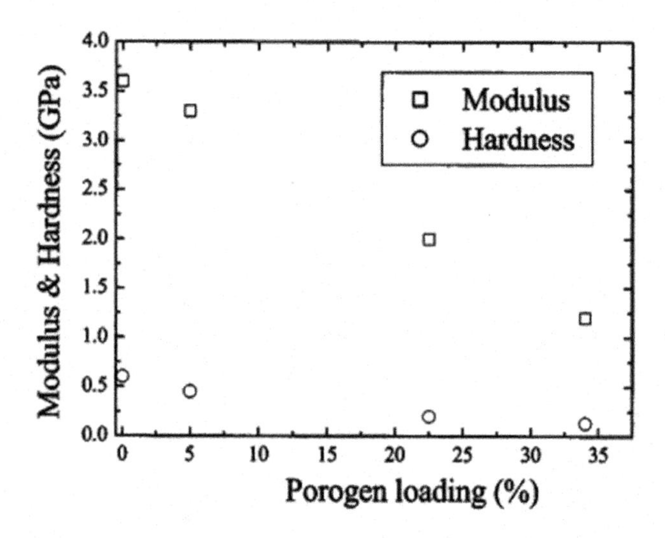

Figure 4.1. *Modulus and hardness variations for an MSQ-based material with subtractive porosity, introduced by various levels of porogen loading [MAE 03]*

A larger porosity also leads to challenges for other processing steps used in the integration. After the low-*k* etching, a metallic barrier has to be deposited in the pattern to prevent copper diffusion. Increased porosity would lead to difficulties in conformal barrier deposition. Alternate processes to physical vapor deposition, such as chemical vapor deposition (CVD) and atomic layer deposition processes, are being investigated [ROS 00]. Increasing the porosity will lead to a deeper penetration of barrier precursors inside the pores that would degrade the porous low-*k* and create conduction paths, as illustrated in Figure 4.2 [POS 08]. The solution to prevent this barrier precursor diffusion is to seal the pore. Different approaches have been proposed in the literature to achieve this, mainly by dense layer deposition, thin layer deposition [MOU 03] or plasma treatments [HOY 04, POS 05a]. Other techniques such as UV treatment have also been proposed [WHE 04].

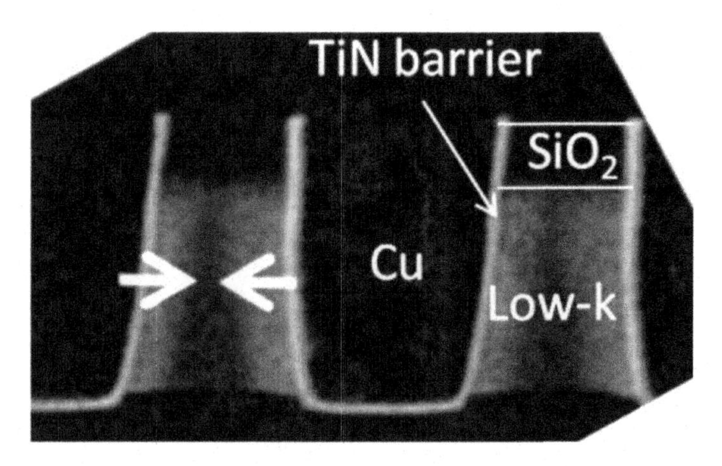

Figure 4.2. *Titanium map measured by Energy Filtered Transmission Electron Microscopy on copper low-k interconnect structure. The TiN barrier is deposited by chemical vapor deposition (CVD) [POS 08]*

Finally, structural dimensions are approaching the sub-10 nm node, and improving the material porosity would increase the pore size. As a result, there is already less than one order of magnitude of difference between the smallest structural dimensions and the pore radii. The difference will reduce with larger porosity. It is obvious that stochastic effects will become important which will lead to performance variability.

4.2. Process solutions for dielectric constant reduction

Several solutions have been proposed in the literature to address porous low-*k* damage during plasma processes. The simplest solution in terms of integration is to work on the plasma processes. In the following, we will describe alternative etching processes and curing processes to avoid plasma-induced damage or to repair the damaged low-*k*.

4.2.1. *Plasma etching with barrier passivation layers*

The solution proposed by Posseme *et al.* consists of using a fluorocarbon-free plasma etching process [POS 13]. In this case, a fluorocarbon-free gas (SiF_4) is used with Ar and N_2 for the etching process. Dense barrier layers form at the sidewalls during the process and help in controlling the pattern profile while preventing the species diffusion through the pore network. The barrier layers are removed after the etching process, leaving only unmodified low-*k* dielectric. As shown in Figure 4.3, acceptable profiles with minimal plasma-induced damage can be obtained by this approach. The challenge with such an approach is to form layers that are effective barriers to species diffusion while being thin-enough to enable pattern profile control. In their chapter, the barrier layers provide pattern profile control and limit plasma species diffusion, resulting in a dielectric constant increase of only 4% after etching [POS 13].

4.2.2. *Limiting species reactivity at lower temperature*

Another approach, proposed by Iacopi *et al.*, consists of using a low wafer temperature to increase the species sticking coefficient and recombination rate. They report the removal of photoresist in O_2-based plasmas at cryogenic temperature and show a threefold reduction of the extent of plasma-induced damage [IAC 11]. When the sample temperature is reduced, the reactive radicals stick at the outer surface and react with plasma species to form stable molecules that are much less reactive than radicals. This results in a strong reduction of plasma-induced damage. Although preliminary results were promising, such processes have been scarcely reported in the literature [STA 13].

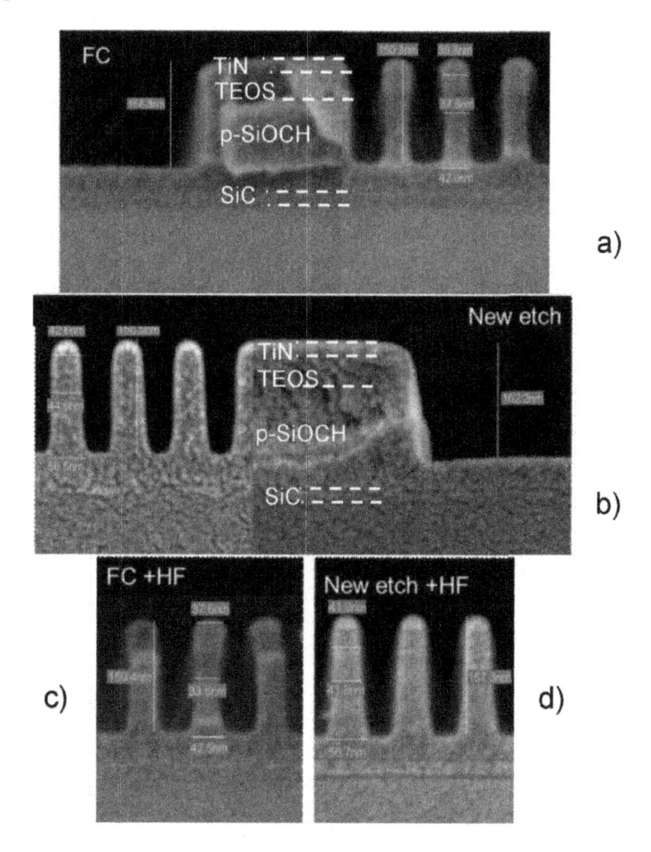

Figure 4.3. *Profile evolution comparison of porous SiOCH (PECVD, porosity 27%, k = 2.3) pattern structure after etching using either conventional FC-based chemistry a) followed by an HF dip c) or new etch chemistry b) followed by an HF dip d) [POS 13]*

4.2.3. *Pore filling using cryogenic plasma*

Another approach of low wafer temperature during plasma etching is to fill the pores during the etching. In addition to the larger sticking coefficient at low temperature, plasma species can condense and diffuse in the pores because of surface tensions [IAC 12]. This fills the pore network and blocks the radicals' diffusion, preventing plasma-induced damage [BAK 13b, ZHA 13]. Cryogenic plasmas have been investigated for years for deep silicon etching [DUS 14]. The concept, described in Figure 4.4, consists of the deposition of SiOF passivation layers at Si patterns sidewalls at low temperature to block the lateral etching. When the sample is heated back to room temperature, such species become volatile and the patterns are passivation layers-free without additional cleaning processes. For porous material etching, condensation and diffusion of etch by-products such as alkyl, alcohols and SiOF leads to pores filling, making the material dense during the etching process [DUS 14]. The condensed species can be easily removed by thermal annealing after the etching process without additional damage [ZHA 13]. This approach is disruptive and highly promising, but requires cryogenic plasma. Such processes require specific tool development and induce high excursion thermal cycles that could be detrimental to complex interconnects structures.

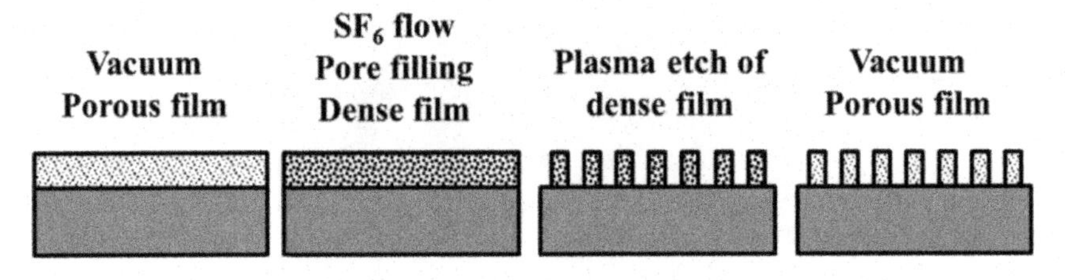

Figure 4.4. *Description of the process flow for the pore filling with cryogenic plasma [IAC 12]*

4.2.4. *Curing processes*

Instead of avoiding plasma-induced damage, we can also work on restoration processes to recover the dielectric properties of the plasma-damaged low-*k*. The dominant approach for porous material repair is silylation. It relies on the grafting of a silicon and carbon containing group on silanol terminations of the porous low-*k*. Many silylation agents have been proposed, including hexamethyldisilazane (HMDS), tetramethyldisilazane (TMDS), trimethylchlorosilane (TMCS), trimethylsilyldiethylamine (TMSDEA), trimethylsilylacetate (OTMSA) and dichlorodimethylsilane (DMDCS) [CHA 02, CHA 07]. Among the candidates for silylation, HMDS is the dominant molecule. After porous low-*k* silylation, silanol terminations are replaced by hydrophobic groups, which prevents moisture uptake [MOR 02]. The main limitation of this approach is the steric hindrance of the large silylation molecules that react first on the outer surface and prevent the deeper diffusion of the silylation agents inside the pores [RAJ 06]. It is thus much more difficult to repair the inner pores which are partially hydrophilic than the outer surface that can be repaired. With current porous low-*k* dielectrics with a pore size below 3 nm and a porosity around 25%, the silylation process is limited to the outer surface of the low-*k*, which explains why this process is not yet used in production. However, with the increase in pore size and interconnectivity at lower dielectric constant, silylation processes will become more efficient which makes them promising for the future.

4.3. Material solutions for dielectric constant reduction

The second approach to lower the dielectric constant while reducing plasma-induced damage is to change the interlayer dielectric material. Indeed, the sensitivity to plasma processes is more or less pronounced depending on the material structure and composition.

4.3.1. *Plasma-resistant porous SiOCH*

The plasma-induced damage originates from species diffusion inside the pores and methyl groups removal. From a material standpoint, the damage can be reduced by either reducing the diffusivity of the plasma radicals or increasing the amount of carbon to be removed [VOL 10]. In this approach, Dubois *et al.* have proposed oxycarbosilane (Si-CH_2-CH_2-Si) and hybrid oxycarbosilane

($Si-CH_2-CH_2-Si$)/methylsilsesquioxane ($Si-CH_3$) material as porous low-*k* dielectrics [VOL 11, VOL12, MAT 12]. They have shown that the $Si-CH_2-CH_2-Si$ bridges improve the mechanical stability of the material [MAT 12]. Such materials present a larger carbon concentration, smaller pore size and lower porosity. A dielectric constant of 2.0 with a porosity of 34% is obtained, which minimizes the risk of plasma-induced damage [DAR 13, BRU 13]. With this approach, the mechanisms of plasma-induced damage are only slowed down, which reduces the extent of plasma-induced damage but does not prevent it. The minimization of the plasma-induced damage originates from the smaller pore size and the larger carbon concentration, but not from the specific chemical structure of the material (carbon bridges) that is destroyed by the high energy ions at the etch front during the plasma etching steps [DAR 13].

4.3.2. *Organic polymer low-k*

Completely different materials such as polymers can also be integrated as low-*k* dielectrics. In the early days of low-*k* dielectrics, SiOCH and carbon-based polymers were considered for interconnects fabrication. SiOCH became the dominant low-*k* dielectric, mostly because of the low compatibility of polymers with copper and metal barriers [VOL 10]. The new challenges related to the low-*k* porosity renewed the interest of the organic materials as potential low-*k* dielectrics [VAN 11, PAN 11a, GU 11, HIR 12]. Numerous polymers have a dielectric constant lower than 2.5. Porosity addition may also further decrease polymers' dielectric constant. Working devices with copper lines isolated by 30 nm of an organic low-*k* with a constant of 2.25 were demonstrated in 2011 by Pantouvaki *et al.* [PAN 11a] (see Figure 4.5). Compared to porous SiOCH, copper cannot diffuse inside organic low-*k*, which allows for integration without a metallic barrier. The copper lines cross-section is thus enlarged and the line resistance is reduced [HIR 12]. Some work is still required to assess the reliability of polymer-based interconnects for replacing standard porous SiOCH even if significant adhesion of polymer with copper and barrier layers has been reported.

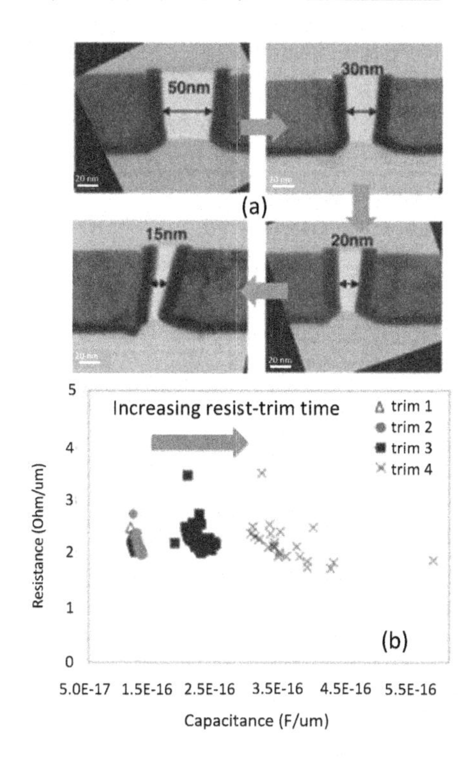

Figure 4.5. *a) TEM images of Cu lines with narrow spacing. The arrows indicate the resist trimming time that controls the dielectric line width. b) RC plots of associated copper lines [PAN 11a]*

4.4. Alternative interconnect architectures for dielectric constant reduction

The last approach to tackle the issue of dielectric constant degradation during interconnects fabrication is to integrate low-*k* dielectrics with alternative integration schemes. Extensive modification of the integration flow is required, but promising solutions to get rid of low-*k* damage during the processing steps can be envisioned.

4.4.1. *Late porogen removal*

Compared to porous materials, dense materials are not significantly damaged by plasma processes since radicals species cannot diffuse inside the material [POS 03]. Integrating low-*k* dielectrics in a dense form and creating the porosity after the damaging steps would thus avoid plasma-induced damage. Shipley proposed this approach under the commercial name "solid-first" in the mid 2000s [ADA 04]. The porous low-*k* dielectric is deposited as a hybrid material consisting of a porous matrix and a sacrificial polymer called porogen (porosity generator). In conventional integration, the porogen is degraded after deposition to release the porosity of the low-*k* matrix. During this thermal step, the low-*k* matrix also crosslinks and stabilizes. In the solid-first integration scheme, the porogen degradation step is shifted either after the plasma etching or after the CMP step to conserve a dense material during all the damaging process steps, as described in Figure 4.6. Since the hybrid material can be seen as a carbon-rich dense SiOCH material, it can be processed using standard plasma etching processes and it is not significantly degraded during the etching steps [EON 07]. Some damage is still observed, however, during hydrogen-based plasma processes because of the large diffusivity of hydrogen radicals even if the pores are filled with the sacrificial porogen [DAR 08a]. With the solid-first integration scheme, Jousseaume *et al.* fabricated single damascene structures with a dielectric constant of 2.2 in 2005 [JOU 05]. This concept is promising to prevent plasma-induced degradation, but low-*k* mechanical deformation during the porogen removal step stresses the interconnects structures and can result in cracking and delamination [DUC 08]. In addition, the top of the copper lines is not protected during the porogen removal, which may lead to copper oxidation if this step is performed after copper polishing [DUC 08].

4.4.2. *Post-porosity plasma protection (P4)*

The concept of solid-first was improved in an innovative scheme proposed by IBM in 2011. To avoid material shrinkage during the porosity generation, Dubois and colleagues proposed to refill the pores of the low-*k* with a sacrificial polymer after the porous low-*k* is stabilized [FRO 11]. This approach, described in Figure 4.7, is named post-porosity plasma protection (P4). Contrary to the solid-first scheme, the porous material is cured before the pores are filled with the sacrificial polymer, which prevents material shrinkage when the protecting polymer is degraded. Frot *et al.* demonstrated that with the P4 scheme, the plasma-induced damage does not scale with the dielectric constant anymore, opening the way to dielectric constants as low as $k = 1.8$ [FRO 12a, FRO 12b]. Instead of suffering from the low-*k* porosity, the P4 integration scheme takes advantage of the species diffusion to refill the pores, which extends the opportunity to reduce the dielectric constant below 2.0.

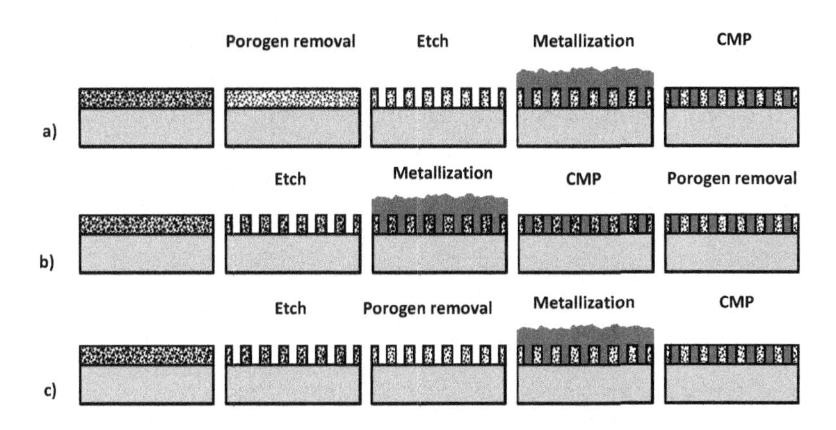

Figure 4.6. *Process flow for a) conventional integration, b) solid-first integration with porosity generation after CMP and c) solid first integration with porosity generation after etching*

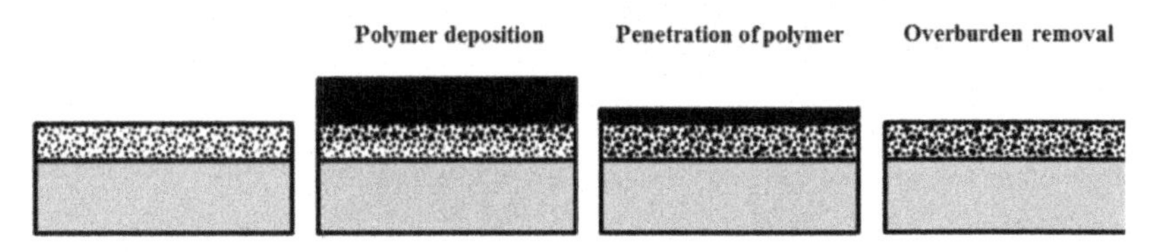

Figure 4.7. *Process flow for the post-porosity plasma protection (P4) approach [FRO 11]*

4.4.3. *Photopatternable low-k dielectrics (PPLK)*

To avoid plasma processes, Lin *et al.* proposed, in 2010, a low-*k* dielectric that is photopatternable [LIN 10a]. Using a lithography step only, they can define structures in the low-*k* dielectric. They reported functional single and dual damascene-based interconnects structures in 2010 (see Figure 4.8) [LIN 10a, LIN 10b]. The patterns being defined by lithography, the plasma processing steps are avoided, and thus the low-*k* is not damaged. In practice, a short etching step is still required to remove the capping and/or anti-reflective coating lying in between the interconnects levels and to connect the two levels. Lin *et al.* have demonstrated the capability of PPLK for complex interconnects structural construction with a dielectric constant of 2.7 and sub-200 nm dimensions [LIN 10b]. These values are usable for the largest levels of interconnects but a lower dielectric constant and a higher resolution material are required for PPLK to become an option for replacement of current integration schemes at the first levels of interconnects. Furthermore, the plasma process needed to etch the capping/anti-reflective coating may induce damage in the PPLK.

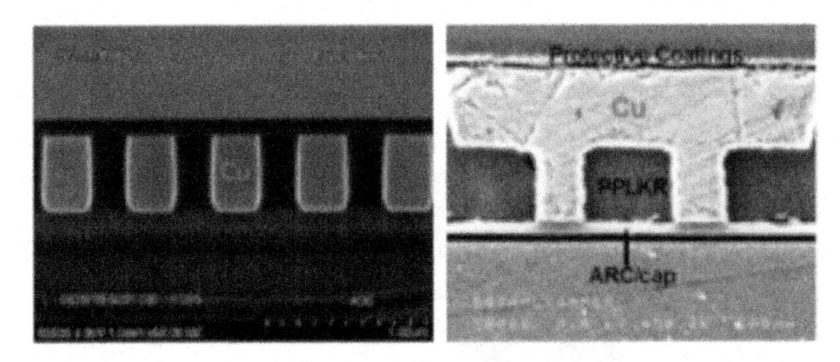

Figure 4.8. *SEM cross-section of single damascene (left-hand side) and dual damascene (right-hand side) interconnects structures formed with PPLK [LIN 10a]*

4.4.4. *Subtractive plasma etching*

The damascene integration was introduced to enable copper integration since copper is difficult to etch. Traditional integration for Al/SiO_2 interconnects was performed by subtractive metal etching, i.e. metal patterning and interspace filling by a dielectric. To avoid low-*k* exposure to plasma processes, it would be possible to drop the damascene integration scheme and move back to the subtractive metal etching. This option presents following advantages:

– the dielectric material is not exposed to plasma processing steps, which avoids plasma-induced damage;

– the metal is deposited as a thin film that increases the grain size and mitigates the size-induced resistivity increase.

The major challenge with subtractive metal etching lies in the metal etching step. Copper etching is difficult because of the low volatility of etch by-products. Halogen-based plasmas require a substrate temperature larger than 200°C [LEE 97]. Recently, Hess and colleagues proposed hydrogen-based copper etching, but such processes still need to be proven on an industrial scale [WU 10, WU 11a, WU 11b]. Other metals could potentially be integrated as a replacement for copper but they need to fulfill patterning capability, low resistivity and low electromigration. In addition, a full refoundation of the process flow needs to be done to enable the formation of barriers at the metal sidewalls and to form vias and lines within one single metallization step, as it is the case with dual damascene. Finally, low-*k* dielectrics must be able to fill narrow trenches without gaps, which excludes plasma-enhanced chemical vapor deposition (PECVD) processes. The CMP step that is required to remove the excess of dielectric may damage the low-*k*.

4.4.5. *Air gaps*

The lowest dielectric constant achievable is 1. For this reason, many researchers have tentatively integrated gaps between the metal lines to form the so-called air gap interconnects. Numerous integration flows have been investigated with either a sacrificial insulator that is decomposed through a porous cap [GAI 06, PAN 08] or trenches etching between the metal lines followed by non-conformal deposition of a dielectric [NOG 05, PON 08, PAN 11b]. Although tremendous efforts have been made to manufacture air gap interconnects, no product has yet been commercially released with these structures. In 2007, IBM announced the fabrication of a fully functional 65 nm node integrated circuit with multilevel air gap interconnects with first shipments expected in 2009 (see Figure 4.9) [IBM 07]. The integration of air gap interconnects has been delayed however and no production is expected in a near future anymore because of the extreme complexity of such architecture. The major drawbacks with air gap interconnects are the following:

– the mechanical stability for air gap structures is poorer than for standard interconnects. Most research groups focus on localized air gap interconnects to prevent the collapse of structures while providing reduced dielectric constant in the most critical lines;

– the interface quality of metal with air is not as good as with dielectric, which reduces the electromigration performance of air gap structures. Voids can nucleate and grow, leading to lines opening and defectivity;

– the dielectric breakdown field is lower for localized air gaps than for conventional interconnects;

– local overheating of the circuits can occur because of the lower thermal conductivity of air gap interconnects.

Despite these limitations, the air gap interconnects are considered the ultimate solution for *k* scaling and remain a target in the roadmap of most integrated circuit manufacturers.

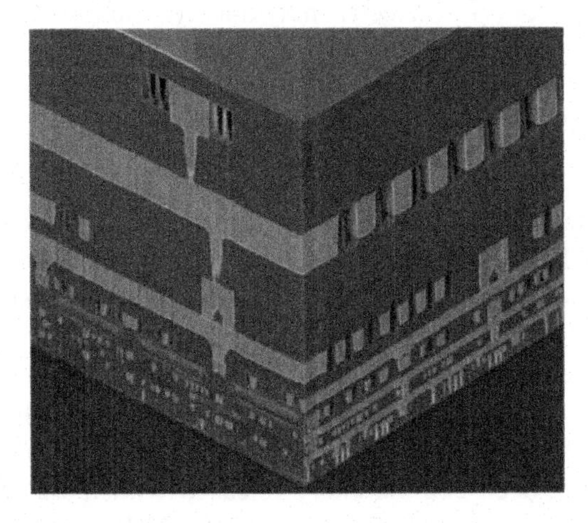

Figure 4.9. *Multilevel integrated circuits featuring air gap interconnects demonstrated by IBM in 2007 [IBM 07]*

4.5. Conclusion

The reduction of the dielectric constant of interconnects insulators is required to improve the circuits performance. The traditional way of increasing porosity is also reaching its limit, with a large porosity to envision integration when using conventional processes. Thus, it is necessary to move from conventional processes to breakthrough innovations either on the processes, materials or integration schemes. Passivating processes, or processes performed at low temperatures, are promising to reduce the diffusion of reactive species and the extent of plasma-induced damage in porous dielectrics. Restoration processes based on silylation are also alternatives to recover the dielectric properties after the integration. From a materials standpoint, porous dielectrics with smaller pore size and larger carbon concentration such as hybrid oligocarbosilanes/methylsilsesquioxane (OCS/MSQ) film are more resistant to plasma damage. Dense or porous organic polymers are also an alternative to porous SiOCH. In order to suppress the plasma-induced damage, low-*k* dielectrics must be dense during the plasma processes, and the porosity must be created after the damaging steps, as proposed by the solid-first approach, or, more promisingly, by the P4 process flow. Photopatterning of the low-*k* is an alternative to plasma processes to create the damascene structures. Finally, air gap interconnects with voids between the metal lines remain the ultimate interconnects structure with a dielectric constant of one, but tremendous effort needs to be done before considering integrating such structures in manufacturing.

Bibliography

[ABD 08] ABDALLAH D.J., MIYAZAKI S., HISHIDA A., *et al.*, "Etching spin-on trilayer masks", *SPIE Advanced Lithography*, vol. 6923, p. 69230, 2008.

[ADA 04] ADAMS T., CALVERT J., GALLAGHER M., *et al.*, Electronic device manufacture, US Patent no. 20040033700, 2004.

[BAI 10] BAILLY F., DAVID T.L., CHEVOLLEAU T., *et al.*, "Roughening of porous SiCOH materials in fluorocarbon plasmas", *Journal of Applied Physics*, vol. 108, no. 1, p. 014906, 2010.

[BAK 06] BAKLANOV M., MAEX K., "Porous low dielectric constant materials for microelectronics", *Philosophical Transactions of the Royal Society A*, vol. 364, no. 1838, pp. 201–215, 2006.

[BAK 13a] BAKLANOV M., DE MARNEFFE J.F., SHAMIRYAN D., *et al.*, "Plasma processing of low-k dielectrics", *Journal of Applied Physics*, vol. 113, no. 4, pp. 041101-1–041101-41, 2013.

[BAK 13b] BAKLANOV M., IACOPI F., VANHAELEMEERSCH S., Protective treatment for porous materials, US Patent no. 8540890, 2013.

[BRU 13] BRUCE R.L., ENGELMANN S., PURUSHOTHAMAN S., *et al.*, "Investigation of plasma etch damage to porous oxycarbosilane ultra low-k dielectric", *Journal of Physics D: Applied Physics*, vol. 46, no. 26, p. 265303, 2013.

[CAR 90] CARDIANUD C., TURBAN G., "Mechanistic studies of the initial-stages of etching of Si and SiO_2 in a CHF_3 plasma", *Applied Surface Science*, vol. 45, no. 2, pp. 109–120, 1990.

[CHA 02] CHANG T.C., LIU P.T., MOR Y.S., *et al.*, "Eliminating dielectric degradation of low-k organosilicate glass by trimethylchlorosilane treatment", *Journal of Vacuum Science Technology B*, vol. 20, no. 4, pp. 1561–1566, 2002.

[CHA 07] CHAABOUNI H., CHAPELON L.L., AIMADEDDINE M., *et al.*, "Sidewall restoration of porous ultra low-k dielectrics for sub-45nm technology nodes", *Microelectronic Engineering*, vol. 84, no. 11, pp. 2595–2599, 2007.

[CHE 07] CHEVOLLEAU T., DARNON M., DAVID T., *et al.*, "Analyses of chamber walls coatings during the patterning of ultra low-k materials with a metallic hard mask: consequences on cleaning strategies", *Journal of Vacuum Science and Technology B*, vol. 25, no. 3, pp. 886–892, 2007.

[DAR 06] DARNON M., CHEVOLLEAU T., EON D., *et al.*, "Etching characteristics of TiN used as hard mask in dielectric etch process", *Journal of Vacuum Science and Technology B*, vol. 24, pp. 2262–2270, 2006.

[DAR 07] DARNON M., CHEVOLLEAU T., JOUBERT O., *et al.*, "Undulation of sub-100nm porous dielectric structures: a mechanical analysis", *Applied Physics Letters*, vol. 91, no. 19, pp. 194103–194103-3, 2007.

[DAR 08a] DARNON M., CHEVOLLEAU T., DAVID T., *et al.*, "Modifications of dielectric films induced by plasma ashing processes: hybrid versus porous SiOCH materials", *Journal of Vacuum Science and Technology B*, vol. 26, no. 6, pp. 1964–1970, 2008.

[DAR 08b] DARNON M., CHEVOLLEAU T., EON D., *et al.*, "Patterning of narrow porous SiOCH trenches using a TiN hard mask", *Microelectronic Engineering*, vol. 85, no. 11, pp. 2226–2235, 2008.

[DAR 10] DARNON M., CHEVOLLEAU T., DAVID T.L., *et al.*, "Patterning of porous SiOCH using an organic mask: comparison with a metallic masking strategy", *Journal of Vacuum Science & Technology B*, vol. 28, no. 1, pp. 149–156, 2010.

[DAR 13a] DARNON M., CASIEZ N., CHEVOLLEAU T., *et al.*, "Impact of low-k structure and porosity on etch processes", *Journal of Vacuum Science and Technology B*, vol. 31, no. 1, pp. 011207–011207-12, 2013.

[DAR 13b] DARNON M., CHEVOLLEAU T., LICITRA C., *et al.*, "Analysis of water adsorption in plasma-damaged porous low-k dielectric by controlled-atmosphere infrared spectroscopy", *Journal of Vacuum Science and Technology B*, vol. 31, no. 6, pp. 061206–061206-6, 2013.

[DEL 99] DELMAS-BENDHIA S., CAIGNER F., SICARD E., *et al.*, "On chip sampling in CMOS integrated circuits", *IEEE Transactions on Electromagnetic Compatibility*, vol. 41, pp. 403–406, 1999.

[DUC 08] DUCOTE J., DAVID T., POSSEME N., *et al.*, "Comparison between hybrid and porous dielectric material (SiOCH) integration strategies for interconnect technologies", *55th International AVS Symposium & Topical Conferences*, 2008.

[DUC 14] DUCOTÉ J., POSSEME N., DAVID T.L., *et al.*, "Prediction of porous dielectric line wiggling phenomenon with metallic hard mask: from simulation to experiment", *Applied Physics Letters*, vol. 104, no. 23, pp. 231603–231603-4, 2014.

[DUS 14] DUSSART R., TILLOCHER T., LEFAUCHEUX P., *et al.*, "Plasma cryogenic etching of silicon from the early days to today's advanced technologies", *Journal of Physics D: Applied Physics*, vol. 47, p. 123001, 2014.

[EON 07] EON D., DARNON M., CHEVOLLEAU T., *et al.*, "Etch mechanisms of hybrid low-k maerial (SiOCH with porogen) in fluorocarbon based plasma", *Journal of Vacuum Science and Technology B*, vol. 25, no. 3, pp. 715–720, 2007.

[FRO 11] FROT T., VOLKSEN W., MAGBITANG T., *et al.*, "Post porosity plasma protection a new approach to integrate k 2.2 porous ULK materials", *IEEE Interconnect Technology Conference and 2011 Materials for Advanced Metallization (IITC/MAM)*, pp. 1–3, 2011.

[FRO 12a] FROT T., VOLKSEN W., MAGBITANG T., *et al.*, "Post porosity plasma protection applied to a wide range of ultra low-k materials", *IEEE Interconnect Technology Conference (IITC)*, pp. 1–3, 2012.

[FRO 12b] FROT T., VOLKSEN W., PURUSHOTHAMAN S., *et al.*, "Post porosity plasma protection: scaling of efficiency with porosity", *Advanced Functional Materials*, vol. 22, no. 14, pp. 3043–3050, 2012.

[FUA 01] FUARD D., JOUBERT O., VALLIER L., *et al.*, "High density plasma etching of low k dielectric polymers in oxygen-based chemistries", *Journal of Vacuum Science and Technology B*, vol. 19, no. 2, pp. 447–455, 2001.

[FUA 03] FUARD D., Study and characterization of plasma etch processes for sub-0.1µm technology node organization: CNRS, PhD Thesis, 2003.

[GAI 06] GAILLARD F., DE PONTCHARRA J., GOSSET L.G., *et al.*, "Chemical etching solutions for air gap formation using a sacrificial oxide/polymer approach", *Microelectronic Engineering*, vol. 83, nos. 11–12, pp. 2309–2313, 2006.

[GAL 89] GALLATIN G.M., ZAROWIN C.B., "Unified approach to the temporal evolution of surface profiles in solid etch and deposition processes", *Journal of Applied Physics*, vol. 65, no. 12, pp. 5078–5088, 1989.

[GRI 03] GRILL A., NEUMAYER D.A., "Structure of low dielectric constant to extreme low dielectric constant SiCOH films: Fourier transform infrared spectroscopy characterization", *Journal of Applied Physics*, vol. 94, no. 10, pp. 6697–6707, 2003.

[GU 11] GU X., NEMOTO T., TOMITA Y., *et al.*, "Cu damascene interconnects with an organic low-k fluorocarbon dielectric deposited by microwave excited plasma enhanced CVD", *IEEE Interconnect Technology Conference and 2011 Materials for Advanced Metallization (IITC/MAM)*, pp. 1–3, 2011.

[HIR 12] HIRAI M., AKIYAMA Y., KOGA K., *et al.*, "Integration of a low-k organic polymer material (k = 2.3) for reducing both resistance and capacitance", *IEEE Interconnect Technology Conference (IITC)*, pp. 1–3, 2012.

[HOO 05] HOOFMAN R.J.O.M., VERHEIJDEN G.J.A.M., MICHELON J., *et al.*, "Challenges in the implementation of low-k dielectrics in the back-end of line", *Microelectronic Engineering*, vol. 80, pp. 337–344, 2005.

[HOW 91] HOWARD J.B., STEINBRUCHEL C., "Reactive ion etching of copper in $SiCl_4$-based plasmas", *Applied Physics Letter*, vol. 59, no. 4, pp. 914–916, 1991.

[HOY 04] HOYAS A.M., SCHUMACHER J., WHELAN C.M., *et al.*, "Plasma sealing of low-k dielectric polymer", *Microelectronic Engineering*, vol. 76, no. 32, pp. 32–37, 2004.

[IAC 04] IACOPI F., BRONGERSMA S.H., VANDEVELDE B., *et al.*, "Challenges for structural stability of ultra-low-k-based interconnects", *Microelectronic Engineering*, vol. 75, no. 1, pp. 54–62, 2004.

[IAC 11] IACOPI F., CHOI J.H., TERASHIMA K., *et al.*, "Cryogenic plasmas for controlled processing of nanoporous materials", *Physical Chemistry Chemical Physics*, vol. 13, pp. 3634–3637, 2011.

[IAC 12] IACOPI F., STAUSS S., TERASHIMA K., *et al.*, "Cryogenic approaches to low-damage patterning of porous low-k films", *Plasma Etch and Strip in Microelectronics Conference*, Grenoble, France 2012.

[IBM 07] IBM, IBM brings nature to computer chip manufacturing, Press Release, May 2007, available at http://www-03.ibm.com/press/us/en/press release/21473.wss.

[JOU 05] JOUSSEAUME V., ASSOUS M., ZENASNI A., *et al.*, "Cu/ulk (k = 2.0) integration for 45 nm node and below using an improved hybrid material with conventional beol processing and a late porogen removal", *IEEE Interconnect Technology Conference (IITC)*, pp. 60–62, 2005.

[JOU 07a] JOUSSEAUME V., FAVENNEC L., ZENASNI A., *et al.*, "Porous ultra-low-k deposited by PECVD: from deposition to material properties", *Surface & Coatings Technology*, vol. 201, pp. 9248–9251, 2007.

[JOU 07b] JOUSSEAUME V., ROLLANDA G., BABONNEAU D., *et al.*, "Structural study of nanoporous ultra low-k dielectrics using complementary techniques: ellipsometric porosimetry, X-ray reflectivity and grazing incidence small-angle X-ray scattering", *Applied Surface Science*, vol. 254, pp. 473–479, 2007.

[KAB 13] KABANSKY A.Y., TAN S.H., HUDSON E.A., *et al.*, "Effective defect control in TiN MHM Cu/low-k DD process", *Electrochemical Society Transactions*, vol. 58, no. 6, pp. 143–150, 2013.

[KEI 03] KEIL D.L., HELMER B.A., LASSIG S., "Review of trench and via plasma etch issues for copper dual damascene in undoped and fluorine-doped silicate glass oxide", *Journal of Vacuum Science and Technology B*, vol. 21, no. 5, pp. 1969–1985, 2003.

[LAZ 05] LAZZERI P., HUA X., OEHRLEIN G.S., *et al.*, "Porosity-induced effects during C_4F_8/90% Ar plasma etching of silica-based ultralow-k dielectrics", *Journal of Vacuum Science Technology B*, vol. 23, no. 4, pp. 1491–1498, 2005.

[LEE 97] LEE S.-K., CHUN S.-S., HWANG C.-Y., *et al.*, "Reactive ion etching mechanism of copper film in chlorine-based electron cyclotron resonance plasma", *Japanese Journal of Applied Physics*, vol. 36, no. 1, pp. 50–55, 1997.

[LEE 10] LEE J., GRAVES D.B., "Synergistic damage effects of vacuum ultraviolet photons and O_2 in SiCOH ultra-low-k dielectric films", *Journal of Physics D: Applied Physics*, vol. 43, no. 42, p. 425201, 2010.

[LIN 10a] LIN Q., CHEN S.T., NELSON A., *et al.*, "Multilevel integration of patternable low-k material into advanced Cu beol", *Proceeding of SPIE Conference*, vol. 7639, pp. 76390J-1–76390J-8, 2010.

[LIN 10b] LIN Q., NELSON A., CHEN S.-T., *et al.*, "Integration of photo-patternable low-k material into advanced Cu back-end-of-the-line", *Japanese Journal of Applied Physics*, vol. 49, no. 5, p. 05FB02, 2010.

[LOU 04] LOUVEAU O., BOURLOT C., MARFOURE A., *et al.*, "Dry ashing process evaluation for porous ULK films", *Microelectronic Engineering*, vol. 73, pp. 351–356, 2004.

[MAE 03] MAEX K., BAKLANOV M.R., SHAMIRYAN D., *et al.*, "Low dielectric constant materials for microelectronics", *Journal of Applied Physics*, vol. 93, no. 11, pp. 8793–8841, 2003.

[MAT 12] MATSUDA Y., RATHORE J.S., INTERRANTE L.V., *et al.*, "Moisture-insensitive polycarbosilane films with superior mechanical properties", *ACS Applied Materials and Interfaces*, vol. 4, no. 5, pp. 2659–2663, 2012.

[MEB 14] MEBARKI M., DARNON M., JENNY C., *et al.*, "Role of the mask on contact etching at the 20 nm node", *Proceedings of the Plasma Etch and Strip in Microelectronics (PESM) Conference*, Grenoble, France, 12–13 May 2014.

[MOO 10] MOON C.S., TAKEDA K., SEKINE M., *et al.*, "Etching characteristics of organic low-k films interpreted by internal parameters employing a combinatorial plasma process in an inductively coupled H_2/N_2 plasma", *Journal of Applied Physics*, vol. 107, no. 11, pp. 113310–113310-8, 2010.

[MOR 02] MOR Y.S., CHANG T.C., LIU P.T., *et al.*, "Effective repair to ultra-low-k dielectric material (k 2.0) by hexamethyldisilazane treatment", *Journal of Vacuum Science Technology B*, vol. 20, no. 4, pp. 1334–1338, 2002.

[MOU 03] MOURIER T., JOUSSEAUME V., FUSALBA F., *et al.*, "Porous low-k pore sealing process study for 65 nm and below technologies", *Interconnect Technology Conference*, p. 245, 2003.

[NOG 05] NOGUCHI J., SATO K., KONISHI N., *et al.*, "Process and reliability of air-gap Cu interconnect using 90 nm node technology", *IEEE Transactions on Electron Devices*, vol. 52, no. 3, pp. 352–359, 2005.

[OHN 98] OHNO K., SATO M., ARITA Y., “Reactive ion etching of copper films in $SiCl_4$ and N_2 mixture”, *Japanese Journal of Applied Physics*, vol. 28, no. 6, pp. 1070–1072, 1989.

[PAN 08] PANTOUVAKI M., HUMBERT A., VANBESIEN E., *et al.*, “Air gap formation by UV-assisted decomposition of CVD material”, *Microelectronic Engineering*, vol. 85, no. 10, pp. 2071–2074, 2008.

[PAN 11a] PANTOUVAKI M., HUFFMAN C., ZHAO L., *et al.*, “Advanced organic polymer for the aggressive scaling of low-k materials”, *Japanese Journal of Applied Physics*, vol. 50, no. 4, p. 04DB01, 2011.

[PAN 11b] PANTOUVAKI M., SEBAAI F., KELLENS K., *et al.*, “Dielectric reliability of 70 nm pitch air-gap interconnect structures”, *Microelectronic Engineering*, vol. 88, no. 7, pp. 1618–1622, 2011.

[PAS 97] PASSEMARD G., FUGIER P., NOEL P., *et al.*, “Study of fluorine stability in fluoro-silicate glass and effects on dielectric properties”, *Microelectronic Engineering*, vol. 33, pp. 335–342, 1997.

[PER 05] PERRET A., CHABERT P., JOLLY J., *et al.*, “Ion energy uniformity in high-frequency capacitive discharges”, *Applied Physics Letter*, vol. 86, pp. 021501-1–021501-3, 2005.

[PON 94] PONS M., PELLETIER J., JOUBERT O., “Anisotropic etching of polymers in SO_2/O_2 plasmas: hypotheses on surface mechanisms”, *Journal of Applied Physics*, vol. 75, no. 9, pp. 4709–4715, 1994.

[PON 08] PONOTH S., HORAK D., NITTA S., *et al.*, “Self-assembly based air-gap integration”, *Electrochemical Society Meeting*, vol. 28, p. 2074, 2008.

[POS 03] POSSEME N., CHEVOLLEAU T., JOUBERT O., *et al.*, “Etching mechanisms of low-k SiOCH and selectivity to SiCH and SiO_2 in fluorocarbon based plasmas”, *Journal of Vacuum Science Technology B*, vol. 21, no. 6, pp. 2432–2440, 2003.

[POS 04] POSSEME N., CHEVOLLEAU T., JOUBERT O., *et al.*, “Etching of porous SiOCH materials in fluorocarbon-based plasmas”, *Journal of Vacuum Science Technology B*, vol. 22, no. 6, pp. 2772–2784, 2004.

[POS 05a] POSSEME N., DAVID T., CHEVOLLEAU T., *et al.*, “A novel low-damage methane-based plasma ash chemistry (CH_4/Ar): limiting metal barrier diffusion into porous low-k materials”, *Electrochemical and Solid-State Letters*, vol. 8, no. 5, pp. G112–G114, 2005.

[POS 05b] POSSEME N., DAVID T., DARNON M., *et al.*, "Hard mask composition and etching chemistry effects on porous ultra low-k material modification", *6th International Conference on Microelectronics and Interfaces*, unpublished, 2005.

[POS 06] POSSEME N., MAURICE C., BRUN PH., *et al.*, "New etch challenges for the 65 nm technology node low-k integration using an enhanced trench first hard mask architecture", *IEEE Interconnect Technology Conference*, pp. 36–38, 2006.

[POS 07] POSSEME N., CHEVOLLEAU T., DAVID T., *et al.*, "Mechanisms of porous dielectric film modification induced by reducing and oxidizing plasmas", *Journal of Vacuum Science and Technology B*, vol. 25, no. 6, pp. 1928–1940, 2007.

[POS 08] POSSEME N., CHEVOLLEAU T., DAVID T.L., *et al.*, "Efficiency of reducing and oxidizing ash plasmas in preventing metallic barrier diffusion into porous SiOCH", *Microelectronic Engineering*, vol. 85, no. 8, pp. 1842–1849, 2008.

[POS 10] POSSEME N., CHEVOLLEAU T., BOUYSSOU R., *et al.*, "Residue growth on metallic hard mask after dielectric etching in fluorocarbon-based plasmas Part I: mechanisms", *Journal of Vacuum Science Technology B*, vol. 28, no. 4, pp. 809–816, 2010.

[POS 11] POSSEME N., BOUYSSOU R., CHEVOLLEAU T., *et al.*, "Residue growth on metallic hard mask after dielectric etching in fluorocarbon based plasmas. II. Solutions", *Journal of Vacuum Science Technology B*, vol. 29, no. 1, pp. 011018–011018-10, 2011.

[POS 13] POSSEME N., VALLIER L., KAO C.L., *et al.*, "New fluorocarbon free chemistry proposed as solution to limit porous sioch film modification during etching", *IEEE Interconnect Technology Conference (IITC)*, pp. 1–3, 2013.

[RAJ 06] RAJAGOPALAN T., LAHLOUH B., LUBGUBAN J.A., *et al.*, "Investigation on hexamethyldisilazane vapor treatment of plasma-damaged nanoporous organosilicate films", *Applied Surface Science*, vol. 252, no. 18, pp. 6323–6331, 2006.

[ROS 00] ROSSNAGEL S.M., SHERMAN A., TURNER F., "Plasma-enhanced atomic layer deposition of Ta and Ti for interconnect diffusion barriers", *Journal of Vacuum Science and Technology B*, vol. 18, no. 4, pp. 2016–2020, 2000.

[ROU 05] ROUESSAC V., FAVENNEC L., RÉMIAT B., *et al.*, “Precursor chemistry for ULK CVD”, *Microelectronic Engineering*, vol. 82, pp. 333–340, 2005.

[SEN 97] SENG W.T., HSIEH Y.T., LIN C.F., *et al.*, “Chemical mechanical polishing and material characteristics of plasma enhanced chemically vapor deposited fluorinated oxide thin films”, *Journal of Electrochemical Society*, vol. 144, no. 3, pp. 1100–1105, 1997.

[SHA 04] SHAMIRYAN D., ABELL T., IACOPI F., *et al.*, “Low-k dielectric materials”, *Materials Today*, vol. 7, no. 1, pp. 34–39, January 2004.

[STA 99] STANDAERT T.E.F.M., MATSUO P.J., ALLEN S.D., *et al.*, “Patterning of fluorine-, hydrogen-, and carbon-containing SiO_2-like low dielectric constant materials in high-density fluorocarbon plasmas: comparison with SiO_2”, *Journal of Vacuum Science and Technology A*, vol. 17, no. 3, pp. 741–748, 1999.

[STA 00] STANDAERT T.E.F.M., JOSEPH E.A., OEHRLEIN G.S., *et al.*, “Etching of xerogel in high-density fluorocarbon plasmas”, *Journal of Vacuum Science and Technology A*, vol. 18, no. 6, pp. 2742–2748, 2000.

[STA 13] STAUSS S., MORI S., MUNEOKA H., *et al.*, “Ashing of photoresists using dielectric barrier discharge cryoplasmas”, *Journal of Vacuum Science and Technology B*, vol. 31, no. 6, pp. 061202–061202-8, 2013.

[TAT 05] TATSUMI T., URATA K., NAGAHATA K., *et al.*, “Quantitative control of etching reactions on various SiOCH materials”, *Journal of Vacuum Science and Technology A*, vol. 23, no. 4, pp. 938–946, 2005.

[TON 03] TONOTANI J., IWAMOTO T., SATO F., *et al.*, “Dry etching characteristics of TiN film using Ar/CHF_3, Ar/Cl_2, and Ar/BCl_3 gas chemistries in an inductively coupled plasma”, *Journal of Vacuum Science Technology B*, vol. 21, no. 5, pp. 2163–2168, 2003.

[TRE 98] TREICHEL H., RUHL G., ANSMANN P., *et al.*, “Low dielectric constant materials for interlayer dielectric”, *Microelectronic Engineering*, vol. 40, pp. 1–19, 1998.

[UCH 08] UCHIDA S., TAKASHIMA S., HORI M., *et al.*, “Plasma damage mechanisms for low-k porous SiOCH films due to radiation, radicals, and ions in the plasma etching process”, *Journal of Applied Physics*, vol. 103, pp. 073303–073303-5, 2008.

[VAN 11] VANSTREELS K., PANTOUVAKI M., FERCHICHI A., *et al.*, "Effect of bake/cure temperature of an advanced organic ultra-low-k material on the interface adhesion strength to metal barriers", *Journal of Applied Physics*, vol. 109, no. 7, pp. 074301-1–074301-8, 2011.

[VOL 10] VOLKSEN W., MILLER R.D., DUBOIS G., "Low dielectric constant materials", *Chemical Reviews*, vol. 110, no. 1, pp. 56–110, 2010.

[VOL 11] VOLKSEN W., MAGBITANG T.P., MILLER R.D., *et al.*, "A manufacturing grade, porous oxycarbosilane spin-on dielectric candidate with k ≤ 2.0", *Journal of the Electrochemical Society*, vol. 158, no. 7, pp. G155–G161, 2011.

[VOL 12] VOLKSEN W., PURUSHOTHAMAN S., DARNON M., *et al.*, "Integration of a manufacturing grade, k = 2.0 spin-on material in a single damascene structure", *ECS Journal of Solid State Science and Technology*, vol. 1, no. 5, pp. N85–N90, 2012.

[WAN 84] WANG J.C.,"Young's modulus of porous materials", *Journal of Materials Science*, vol. 19, no. 3, pp. 801–808, 1984.

[WHE 04] WHELAN C.M., TOAN LE Q., CECCHET F., *et al.*, "Sealing of porous low-k dielectric" *Electrochemical and Solid-State Letter*, vol. 7, no. 2, pp. F8–F10, 2004.

[WOR 05] WORSLEY M.A., BENT S.F., GATES S.M., *et al.*, "Effect of plasma interactions with low-k films as a function of porosity, plasma chemistry, and temperature", *Journal of Vacuum Science and Technology B*, vol. 23, no. 2, pp. 395–405, 2005.

[WU 10] WU F., LEVITIN G., HESS D.W., "Low-temperature etching of Cu by hydrogen-based plasmas", *ACS Applied Materials and Interfaces*, vol. 2, pp. 2175–2179, 2010.

[WU 11a] WU F., HESS D.W., LEVITIN G., Low temperature metal etching and patterning, US Patent no. 8679359, 2011.

[WU 11b] WU F., LEVITIN G., HESS D.W., "Mechanistic considerations of low temperature hydrogen-based plasma etching of Cu", *Journal of Vacuum Science and Technology B*, vol. 29, no. 1, pp. 011013–011013-7, 2011.

[YAM 00] YAMASHITA K., ODANAKA S., "Interconnect scaling scenario using a chip level interconnect model", *IEEE Transactions on Electron Devices*, vol. 47, pp. 90–96, 2000.

[YIN 06] YIN Y., RASGON S., SAWIN H.H., "Investigation of surface roughening of low-k films during etching using fluorocarbon plasma beams", *Journal of Vacuum Science and Technology B*, vol. 24, no. 5, pp. 2360–2371, 2006.

[YIN 07] YIN Y., SAWIN H.H., "Impact of etching kinetics on the roughening of thermal SiO_2 and low-k dielectric coral films in fluorocarbon plasmas", *Journal of Vacuum Science and Technology A*, vol. 25, no. 4, pp. 802–811, 2007.

[YIN 08] YIN Y., SAWIN H.H., "Surface roughening of silicon, thermal silicon dioxide, and low-k dielectric coral films in argon plasma", *Journal of Vacuum Science and Technology A*, vol. 26, no. 1, pp. 151–160, 2008.

[ZHA 13] ZHANG L., LJAZOULI R., LEFAUCHEUX P., *et al.*, "Low damage cryogenic etching of porous organosilicate low-k materials using $SF_6/O_2/SiF_4$", *ECS Journal of Solid State Science and Technology*, vol. 2, no. 6, pp. N131–N139, 2013.

List of Authors

Thierry CHEVOLLEAU
CNRS-LTM
Grenoble
France

Maxime DARNON
CNRS-LTM
Grenoble
France

Thibaut DAVID
CEA-LETI-Minatec
Grenoble
France

Nicolas POSSEME
CEA-LETI-Minatec
Grenoble
France

Index

I, L, M, O

P, S, T

T, U, Y

Printed in the United States
By Bookmasters